Foreword

SOME EXPLORATORY ENTERPRISES start with fanfare and end with a quiet burial; some start with hardly a notice, yet end up significantly advancing mankind's knowledge. The Interplanetary Pioneers more closely fit the latter description. When the National Aeronautics and Space Administration started the program a decade ago it received little public attention. Yet the four spacecraft, designated Pioneers 6, 7, 8, and 9, have faithfully lived up to their name as defined by Webster, "to discover or explore in advance of others." These pioneering spacecraft were the first to systematically orbit the Sun at widely separated points in space, collecting information on conditions far from the Earth's disturbing influence. From them we have learned much about space, the solar wind, and the fluctuating bursts of cosmic radiation of both solar and galactic origin.

These Pioneers have proven to be superbly reliable scientific explorers, sending back information far in excess of their design lifetimes over a period that covers much of the solar cycle.

This publication attempts to assemble a full accounting of this remarkable program. Written by William R. Corliss, under contract with NASA, it is organized as Volume I: Summary (NASA SP-278); Volume II: System Design and Development (NASA SP-279); and Volume III: Operations and Scientific Results (NASA SP-280). In a sense it is necessarily incomplete, for until the last of these remote and faithful sentinels falls silent, the final word is not at hand.

HANS MARK
Director
Ames Research Center
National Aeronautics and
Space Administration

Contents

	Page
Chapter 1. PIONEER OPERATIONS	1
Chapter 2. PRELAUNCH ACTIVITIES	3
Facilities Involved in Pioneer Prelaunch Activities	3
Organizations and Their Responsibilities	4
Significant Events in the Pioneer Prelaunch Phase	9
Prelaunch Schedules—Planned and Actual	13
The Actual Prelaunch Phases and How They Compared	17
Chapter 3. LAUNCH TO DSS ACQUISITION	25
Performance of the Delta Launch Vehicle	26
Tracking and Data Acquisition	36
Spacecraft Performance	44
Chapter 4. FROM DSS ACQUISITION TO THE BEGINNING OF THE CRUISE PHASE	45
Sequence of Events	45
Pioneer Operations—Acquisition to Cruise Phase	49
CHAPTER 5. SPACECRAFT PERFORMANCE DURING THE CRUISE PHASE	55
Pioneer-6 Performance	55
Pioneer-7 Performance	68
Pioneer-8 Performance	73
Pioneer-9 Performance	74
References	75
Chapter 6. PIONEER SCIENTIFIC RESULTS	77
The Goddard Magnetic Field Experiment (Pioneers 6, 7, and 8)	78
The MIT Plasma Probe (Pioneers 6 and 7)	85
The Ames Plasma Probe (All Pioneers)	90
The Chicago Cosmic-Ray Experiment (Pioneers 6 and 7)	97
The GRCSW Cosmic-Ray Experiment (All Pioneers)	102
The Minnesota Cosmic-Ray Experiment (Pioneers 8 and 9)	111
The Stanford Radio Propagation Experiment (All Pioneers)	115
Radio Propagation Experiments Using the Spacecraft Carrier (All Pioneers)	124

	Page
TRW Systems Electric Field Experiment (Pioneers 8 and 9)	127
The Goddard Cosmic Dust Measurements (Pioneers 8 and 9)	135
The Pioneer Celestial Mechanics Experiment (All Pioneers)	138
Solar Weather Monitoring	141
References	142
BIBLIOGRAPHY	147
INDEX	151

CHAPTER 1

Pioneer Operations

THIS VOLUME DESCRIBES the long chain of events that began with the arrival, checkout, and launch of the Pioneer spacecraft at Cape Kennedy and culminated in the publication of results in the scientific journals. There were five major links in the chain, each beginning and ending with a critical event:

Phase 1. *Prelaunch Operations*—Began with the arrival of the spacecraft at the Cape and ended with the launch

Phase 2. *Launch to DSS Acquisition*—Spacecraft usually acquired first by Deep Space Station (DSS) at Johannesburg

Phase 3. *Near-Earth Operations*—Commenced with DSS acquisition and ended with completion of all orientation maneuvers

Phase 4. *Nominal and Extended Cruise*—From completion of orientation maneuvers to end of useful spacecraft life

Phase 5. *Presentation of Scientific and Engineering Results*—Began as soon as the scientific instruments were turned on and ended only when the data became superseded

Scientific data may remain viable for decades, with information of value still being extracted after the spacecraft itself has stopped transmitting.

The operational histories of the five Pioneer spacecraft could be related separately, although this would result in five highly repetitive chapters, and the comparison of spacecraft performance and cooperative spacecraft activities would be difficult. The descriptions of spacecraft operations, therefore, are organized with a chapter assigned to each of the five phases established above.

A Pioneer launch required the coordination of thousands of people located not only at the launch site but also at the tracking stations around the world and at the communication focal points at the Jet Propulsion Laboratory's Space Flight Operations Facility (SFOF) and the Ames Research Center. Some measure of organization had to be superimposed on these people and the operations they performed with the Pioneer spacecraft, the Delta launch vehicle, and the Deep Space Network. Ames Research Center, as the overall program manager, established principles and general specifications for all operational phases. The two most significant of the Ames specifications were:

(1) PC-046—Pioneer Flight Operations (for Block I)
(2) PC-146—Pioneer Space Flight Operations (for Block II)

These specifications spelled out—in some 6 in. of documentation for each block of Pioneers—the operational requirements for each facility and the many participating organizations. Also defined were critical technical terms and the interfaces between the organizations involved. Mutual understanding of what was to be done, and by whom, were the goals of these documents. In addition, they were supplemented by separate Ames operations specifications for each mission as well as by more detailed definitions of critical organizational interfaces.

In themselves, the Ames specifications were too broad to provide the second-by-second directions required by the launch teams at the Cape from Ames, Goddard, JPL, the Air Force, McDonnell-Douglas, TRW Systems, and other participating organizations. After a spacecraft was launched successfully, day-by-day instructions were needed by DSS crews working the spacecraft. At the working level then, a host of countdown documents, operations directives to downrange stations, and other schedules translated the broad, general Ames specifications into working documents.

Documentation seems a rather dull aspect of such a fascinating topic as the scientific exploration of deep space. Spacecraft, however, are launched and operated by people, even though machines (particularly computers) now make more and more of the routine decisions. People work in concert only when they all understand an activity in the same terms and work from a detailed schedule. Without question a major accomplishment of the space program has been its blending of people and equipment into a smoothly functioning global apparatus. The Pioneer program has been an effective part of this larger machine.

CHAPTER 2

Prelaunch Activities

THE SUCCESSFUL COMPLETION of spacecraft preship review signaled the beginning of prelaunch activities. The spacecraft was carefully packed and shipped to Cape Kennedy by air. Its arrival at the Cape initiated a 6- to 10-week series of tests and checkout procedures to assure the readiness of the spacecraft and its compatibility with the Delta launch vehicle, the Deep Space Network (DSN), and equipment along the Air Force's Eastern Test Range. If all went well, the pieces did fit together, and the spacecraft was launched successfully. Of course, the real picture was more complicated.

Basically, one must view Cape Kennedy as a production line where spacecraft and launch vehicles meet, are tested, and fired. Such a complex enterprise requires rigorous scheduling and definition of responsibilities, leading, in this case, to the launches of the Pioneer spacecraft within their narrow launch windows.

FACILITIES INVOLVED IN PIONEER PRELAUNCH ACTIVITIES

More people and facilities participated during the Pioneer prelaunch and launch activities than at any other time in the mission. Although Cape Kennedy and the Eastern Test Range's downrange stations were the focal points during this phase of operations, the DSN and SFOF were all involved in various tests and during various checkout procedures. As the moment of launch approached, more and more of the NASA and Air Force general-purpose facilities "came on the line" for the launch. Radars, optical instrumentation, and telemetry antennas at the Cape and downrange were all, in effect, waiting for the Delta and its Pioneer payload during the minutes before launch. Likewise, critical antennas at some of the DSN's Deep Space Stations broke off from tracking Mariners and Pioneers already out in space and swung toward the points where the new Pioneer was expected to come over the horizon.

The major facilities concerned with a Pioneer launch are described in some detail in Vol. II. Here, only the major functions are reiterated.

Cape Kennedy.—The Cape provided facilities for spacecraft tests, checkout, and integration. Facilities were also provided for mating of spacecraft with launch vehicle and for launch vehicle assembly and launch. The Pioneer Electrical Ground Support Equipment (EGSE) provided an interface between the spacecraft and the launch pad environment. Figure 2–1 shows Launch Complex 17, from which all five Pioneers were launched. The AE, AM, AO, and M buildings are seen in the aerial view presented in figure 2–2. Figure 2–3 was taken from inside the Mission Director's Center.

USAF Eastern Test Range (ETR)—This facility provided tracking and data acquisition services from launch through nominal DSS acquisition at Johannesburg.

The Deep Space Network (DSN)—The DSN supplied tracking, data acquisition, and transmission of command signals to the spacecraft. Included in the DSN is the Deep Space Instrumentation Facility (DSIF) which encompasses all of the DSSs, the SFOF, and the Ground Communications Facility (GCF). For further information see Ch. 8, Vol. II. The Pioneer Ground Operational Equipment (GOE) at selected DSS stations provided an interface between the spacecraft and the generalized DSS equipment.

ORGANIZATIONS AND THEIR RESPONSIBILITIES

Hundreds of people from government and industry applied their talents and training during the testing and checkouts that led to a Pioneer launch. They operated the facilities listed above, or they were part of the launch crews associated with the Delta and the spacecraft.

FIGURE 2–1.—Aerial view of Launch Complex 17 at Cape Kennedy.

FIGURE 2-2.—Aerial view of part of the Cape Kennedy complex. The buildings in the foreground are, from left to right, M, AO, AM, and AE.

FIGURE 2-3.—The Mission Director's Center at Launch Complex 17.

They also had to be given directions and schedules. Ames Research Center, as the arm of NASA managing the Pioneer Program, provided both. The Pioneer Flight Operations Specifications issued by Ames[1] established the prelaunch-phase responsibilities as follows.

Ames Research Center Prelaunch Responsibilities

1. Plan and document the Pioneer space flight operations.
2. Prepare and document a Pioneer space flight operations test plan.
3. Plan and schedule acceptance, integration, and operational readiness tests.
4. Determine requirements and initiate tasks and procurements required to provide the necessary aids and materials used during tests, such as tapes containing simulated spacecraft telemetry and tracking data.
5. Participate in equipment preparation, acceptance, integration, and operational readiness tests.
6. Direct conduct of, monitor, and review acceptance, integration, and operational readiness tests.
7. Prepare and update procedures for mission-dependent activities to be performed during the flight operations.
8. Plan for and initiate tasks and procurements required to provide the necessary GOE, the communications net between the stations supporting the mission, the mission-dependent displays, and the off-line data-processing system.
9. Develop procedures for flight operations associated with mission-dependent equipment for handling on-line and off-line data, for the analysis of spacecraft, experiment, and GOE performance, and for disseminating information to the spacecraft contractor and experimenters.

Jet Propulsion Laboratory Prelaunch Responsibilities

1. Manage and coordinate activities of DSN.
2. Provide personnel to operate mission-independent equipment and Pioneer GOE (except during Type-II orientation) during acceptance, integration, and operational readiness tests, and during flight operations at DSIF.
3. Provide personnel to operate mission-independent equipment during acceptance, integration, and operational readiness tests at SFOF.

---

[1] The basic documents are Pioneer Specifications PC–046, for Block-I Pioneers, and PC–146, for Block-II Pioneers. See Bibliography in Vol. II. Many of these activities, particularly those involving planning, occurred prior to the shipment of the spacecraft to the Cape. The actual equipment preparations and tests prior to launch are covered in detail in the next section.

4. Coordinate communications at SFOF to and from the Pioneer Mission Operations Area and to and from Ames, TRW Systems,[2] and Stanford.

5. Provide and maintain mission-independent equipment at DSIF and SFOF required to support the Pioneer mission.

6. Maintain and repair Pioneer GOE at DSIF sites.

7. Prepare space flight operations plan for use by DSN personnel.

8. Prepare Tracking Instruction Manual.

9. Prepare date as required pertinent to DSN operations for use in the preparation of procedures and training aids to be used during acceptance, integration, and operational readiness tests.

10. Provide data as required pertinent to DSN operations for inclusion in Ames operational documents.

11. Prepare and maintain the Mission Control Center in SFOF to support the Pioneer Project during tests simulating operations from launch to the completion of Type-II orientation maneuvers.

12. Participate in the review of acceptance, integration, and operational readiness tests.

TRW Systems Prelaunch Responsibilities

1. Provide data for and review plans and procedures prepared by Ames for equipment preparation, acceptance, integration, and operational readiness tests and flight operations.

2. Provide data for and review specifications prepared by Ames for test aids, such as magnetic tapes and teletype paper tapes containing simulated spacecraft telemetry data.

3. Provide test aids, such as magnetic tapes and teletype paper tapes containing simulated spacecraft telemetry data.

4. Provide two field engineers at DSS–12 and SFOF to aid Ames in training DSN personnel in operational procedures. Participate in reviewing the telemetry data and command system acceptance tests.

5. Provide as required systems and subsystem engineers at SFOF, knowledgeable in the preparation and operations of the spacecraft and GOE, to participate in acceptance, integration, and operational readiness tests.

6. Provide four field engineers at DSS–12, knowledgeable in the operations of the spacecraft telecommunications and orientation subsystems, the GOE, and the flight dynamics of the spacecraft, to participate in all tests in which the Type-II orientation maneuver is exercised.

---

[2] Spacecraft contractor.

Experimenters' Prelaunch Responsibilities

1. Provide data for plans and procedures prepared by Ames for experiment preparation, acceptance, integration, operational readiness tests, and flight operations.

2. Provide requirements for use by Ames in the preparation of weekly and daily operations plans and in the conduct of the mission as it pertains to the scientific instruments.

3. Stanford must provide a field crew as required to operate the transmitter at Stanford University during equipment preparation, acceptance, integration, operational readiness tests and during flight operations of the Pioneer spacecraft.

NASA Headquarters Prelaunch Responsibilities

1. Review operations readiness of the Pioneer Project prior to launch.
2. Advise the Public Information Office on procedures to be followed for Pioneer.
3. Participate in launch operations at ETR.
4. Participate in flight operations at SFOF.

Goddard Space Flight Center Prelaunch Responsibilities

1. Prepare and submit the Pioneer operations requirements to ETR.

Kennedy Space Center Prelaunch Responsibilities

1. Plan the launch operations.
2. Provide technical direction and implementation of the launch operation.
3. Coordinate activities between NASA, contractors, and ETR groups.

McDonnell-Douglas [3] Prelaunch Responsibilities

1. Prepare planning, reference, and predictive powered-flight trajectories.
2. Review technical documents that relate to launch operations and that have been prepared by other elements within the Pioneer Program.
3. Prepare launch countdown documentation.

Eastern Test Range (USAF) Prelaunch Responsibilities

1. Review technical requirements and documents that relate to launch operations and that have been prepared by other elements within the Pioneer Program.

---

[3] Delta contractor.

2. Provide crews as required to operate ETR stations supporting the Pioneer Program during integration and operational readiness tests and during the launch countdown and powered-flight phase of the Pioneer mission.

From these assignments of responsibility came detailed schedules of procedures telling individuals in all organizations involved what they should do and when. Although the proliferation of plans, task assignments, and schedules may seem overly complex, it is this kind of paperwork that permits large groups of people from diverse organizations to function successfully.

SIGNIFICANT EVENTS IN THE PIONEER PRELAUNCH PHASE

The prelaunch phase of activities consisted of many hundreds of separate items and events; so many, in fact, that the checkout and countdown lists were printed by computers. In addition to the extensive planning activities just described, two other groups of processes and events stand out as important:

(1) Training in operational procedures
(2) Preparation and testing of the spacecraft, launch vehicle, and other mission-dependent hardware

Training in operational procedures was most important during the preparations for the launch of Pioneer A in 1965, when the Pioneer Program was new to ETR and DSN personnel. The Delta, of course, was a familiar sight at the Cape; and the ETR and DSN had already handled spacecraft more complex than the Pioneers. Some of the "different" aspects of the Pioneer launches were:

(1) The unusual orientation maneuvers following launch
(2) The narrow launch window associated with injecting the spacecraft into an orbit roughly parallel to the plane of the ecliptic
(3) The ejection of the Test and Training Satellites (TETR) from Block-II Pioneers
(4) The occultations and flights through the Earth's magnetic tail

The orientation maneuvers, especially, required careful training at the Goldstone DSS site and, in the case of Pioneers 6 and 9, at Johannesburg and Goldstone, respectively, where "partial" Type-II orientation maneuvers were carried out.

The prelaunch preparation and testing of the spacecraft, launch vehicle, and associated hardware commenced with the arrival of the spacecraft at the Cape. These highly important checks and double-checks were performed primarily by Ames, TRW, Goddard, and Kennedy personnel. Although the actual operations varied slightly from mission to mission, the following list of major tasks is representative.[4]

Pioneer Prelaunch Tasks

Task 1. Receipt, unpacking, and inventorying of spacecraft and associated equipment at hangar

Task 2. Verification of mechanical condition of the spacecraft, ground handling equipment, and EGSE

Task 3. Validation of EGSE

Task 4. Spacecraft pneumatic system leak test

Task 5. Spacecraft alignment checks

Task 6. Solar-array performance test

Task 7. Performance checks of critical unit parameters not accessible during an Integrated Systems Test (IST)

Task 8. Integrated System Test (see discussion below)

Task 9. Preparation of spacecraft for mating with third stage

Task 10. Mating of spacecraft to inert third stage

Task 11. Installation of EGSE in blockhouse and its validation

Task 12. Mating of spacecraft and third stage to rest of launch vehicle at the launch pad

Task 13. Preliminary spacecraft on-stand electrical and radio-frequency tests

Task 14. Verification of spacecraft/launch vehicle compatibility with the range

Task 15. Nose fairing fit check

Task 16. Preliminary on-stand Integrated Systems Test

Task 17. Flight readiness demonstration with a spacecraft/launch-vehicle practice countdown

Task 18. Replacement of inert third stage with a live third stage and final spacecraft preparations

Task 19. Verification that the experiments are operating to the satisfaction of the experimenters

Task 20. Spacecraft radio-frequency subsystem test

Task 21. Integrated Systems Test

Task 22. Final spacecraft check prior to dynamic balancing (entails moving spacecraft back to hangar)

Task 23. Spacecraft dynamic balance check

Task 24. Mating of spacecraft with live third stage

Task 25. Dynamic balance test of spacecraft and third stage

Task 26. Reinstallation of EGSE in blockhouse and revalidation

---

[4] TRW Space Technology Laboratories: Test Program Plan, Pioneer Spacecraft Program. Aug. 1964.

Task 27. Mating of spacecraft and live third stage to launch vehicle on pad

Task 28. Spacecraft on-stand electrical and radio-frequency tests

Task 29. Final launch vehicle/spacecraft/range radio-frequency compatibility test

Task 30. Final on-stand Integrated Systems Test

Task 31. Preparation of spacecraft for pre-terminal count (includes installation of live pyrotechnics)

Task 32. Perform joint launch vehicle/spacecraft/range pre-terminal count

Task 33. Terminal count and launch

The Integrated Systems Test, or IST, was performed at least twice for each spacecraft at the Cape. This test (actually a check for launch readiness) was described in Ch. 6, Vol. II. Each spacecraft was subjected to at least one IST before it left the TRW Systems plant for the Cape. A successful IST demonstrated that the spacecraft met all spacecraft performance requirements. It provided a baseline upon which to gage spacecraft operational condition—a background against which to spot trends. It was because of this diagnostic value that the IST was repeated twice or more before launch. The final IST was "on-stand;" that is, carried out when the spacecraft was mated to the Delta rocket on the launch pad. The on-stand IST was the final comprehensive spacecraft check before launch. Recapitulating the IST description in Vol. II, the IST was as close to a realistic operational test as one could get prior to launch and yet be independent of the Delta, the ETR, and the SFOF. A minimum of hardlines were used; radio links were used instead to simulate actual communication links (fig. 2–4). Sun-sensor pulses were also simulated. Basically, a successful IST was a vote of confidence in the spacecraft, even though interfaces with the Delta and DSN were not tested.

The operational readiness tests were dress rehearsals that demonstrated that all personnel, equipment, and facilities participating in a Pioneer launch were ready to support the mission. While the IST was a spacecraft test, the operational readiness test encompassed the entire Pioneer supersystem; that is, the spacecraft and its instruments, the Delta, and the DSN (fig. 2–5). The events and tasks performed and simulated were representative of the actual mission events. The most critical "dry runs" were those simulating the Type-II orientation maneuver in conjunction with the Goldstone DSS. Another simulated situation brought Stanford University into the system, permitting operators there to practice with the spacecraft and ground equipment under realistic conditions. Of course, the normal mission was simulated too, via ersatz commands and telemetry sent over NASA's worldwide communication system (NASCOM) and along ETR communication channels.

FIGURE 2-4.—Pioneer E wired for on-stand checkout at the Cape.

Two operational readiness tests were planned for each prelaunch phase. With Pioneer 6, for example, the first operational readiness test was scheduled for 48-hr duration, with 4 hr devoted to Cape and ETR activities, 13 hr for first-pass events at Johannesburg, and 9 hr for the Type-II orientation maneuver commanded from Goldstone. For Pioneers 6 and 9 the partial Type-II orientation maneuver was simulated at Johannesburg. Normal cruise operations were simulated at all stations. The second operational readiness test just before liftoff was a

FIGURE 2–5.—Scene in control room at SFOF in Pasadena, Calif., during Pioneer-B operational readiness test.

repeat of the first, except that everything was to be compressed into a 24-hr period. If all systems passed the second operational readiness test, a launch readiness review was held by the Pioneer Mission Director (fig. 2–6). The "go/no-go" decision was made at this final meeting. If the decision was "go," the actual countdown began.

PRELAUNCH SCHEDULES—PLANNED AND ACTUAL

The Pioneer spacecraft arrived at the Cape 6 to 10 weeks prior to the planned launch. As the various tests were successfully passed, events multiplied crescendo-like as the day of launch approached. $F-2$, $F-1$, and $F-0$ days were filled with critical tests. The Pioneer project prepared schedules to lend some order to these events. The first "working" schedule of importance was the Detailed Task Sequence prepared by Ames Research Center a few months before the spacecraft was shipped to Cape Kennedy. The Detailed Task Sequence was published as a Pioneer specification. In the case of Pioneer D, which is used as an example here, Specification PC–153 contains the Detailed Task Sequence shown in table 2–1. Although the Detailed Task Sequence was presented

NATIONAL AERONAUTICS AND SPACE ADMINISTRATION

AMES RESEARCH CENTER MOFFETT FIELD, CALIFORNIA

Mission Readiness Review
Pioneer C

Date: December 8, 1967
Place: E & O Building Conference Room 116
Time: 0930 EST.
Chairman: Charles F. Hall, ARC Pioneer Project Manager

AGENDA

Time, EST.	Item	
0930	Introduction	C. F. Hall/ARC
0945	Mission Objectives	C. F. Hall/ARC
1000	Launch Operations Status	J. Nielon/ULO
1020	Launch Vehicle Status	W. McCall/ULO
1100	COFFEE BREAK	
1115	Launch Vehicle/TTS Status	J. Tomasello/ T. Longo/GSFC
1135	TTS Spacecraft Status	P. Burr/GSFC
1205	LUNCH	
1305	Summary of Pioneer Cape Activities	R. W. Holtzclaw/ARC
1325	Pioneer Spacecraft Status	B. O'Brien/TRW
1410	Pioneer Instrument Status	J. Lepetich/ARC
1440	COFFEE BREAK	
1455	Pioneer Post Launch Activities	C. F. Hall/ARC
1515	T & DS Support	J. Thatcher/JPL
	ETR	R. Norman/ULO
	MSFN	D. Bonnell/GSFC
	DSN	J. Thatcher/JPL
	NASCOM	J. Thatcher/JPL
1600	Interstation Conference Station Reports	J. Thatcher/JPL
1645	Summary of Mission Status	C. F. Hall/ARC
1705	Pioneer Program Office Comment	J. Mitchell/ M. Aucremanne/HQ

FIGURE 2-6.—Reproduction of the schedule for the Pioneer-C Mission Readiness Review held at the Cape.

on a time base and was much more detailed than the general Block-II specifications, PC–146, it was not a working schedule in the sense that it specified who, what, when, and where.

The Detailed Task Sequence was next rendered into more specific schedules. It is impractical to reproduce the item-by-item details, but the reader can get a "feel" of these working-level schedules from the Pioneer-D F — 2, F — 1, and F — 0 day schedules in tables 2–1 and 2–2. Detailed descriptions of the tasks to be performed—in terms of switches to be thrown, meters to be read, calibrations to be made, etc.—had to accompany these schedules.

PRELAUNCH ACTIVITIES 15

TABLE 2-1.—*Planned Pioneer-D Typical Detailed Task Sequence* *

Date	Location	Task
Oct. 1	SFOF	Flight path analysis and command group acceptance test
Oct. 1	SFOF	Spacecraft performance analysis area/science analysis and command group acceptance test
Oct. 2	SFOF	SFOF integration test
Oct. 9	ETR	DSS hangar and on-stand compatability test
Oct. 16	ETR	Practice on-stand IST
Oct. 24	SFOF	First operational readiness test
Oct. 25	ETR	Preliminary electrical and radio-frequency checks
Oct. 28	ETR	DSS hangar and on-stand compatability test
Oct. 31	SFOF	Second operational readiness test
Oct. 31	ETR	Practice countdown (pre-fairing installation)
Nov. 1	ETR	Practice countdown (post-fairing installation)
Nov. 4 (major F – 2 day milestones)		
0725 EST	ETR	Countdown initiation
0730	Delta	Task 2, engine checks
0730	Spacecraft	Task I, preparations and spacecraft checks
0900	Delta	Task 3, electrical systems checks
1120	Spacecraft	Task II, pneumatic pressure and fill valve lead check
1210	Delta	Task 5, stray voltage checks
1320	Spacecraft	Task III, service magnetometer
1440	Delta	Task 6A, class B ordnance installation and hookup
1440	Spacecraft	Ordnance installation
1710	Delta	Task 6B, squib installation
1840	ETR	Built-in hold (8 hr, 15 min)
Nov. 5 (major F – 1 day milestones)		
0700 EST	Spacecraft	Task V, remove Red Tag items and protective covers
0725	ETR	Countdown initiation
0730	Delta	Task 7B, second-stage final preparations
0730	Delta	Task 9A, Second-stage propellant servicing setup
0830	Delta	Task 7A, fairing erection
1130	Delta	Task 6C, FW-4 (third stage) hookup
1200	Delta	Task 7B, fairing installation
1330	Spacecraft	Task III, umbilical checks
1400	Delta	Task 9B, second-stage propellant servicing
1730	Delta	Task 7C, blast-band installation
1745	Delta	Task 10, first-stage fueling
1900	Delta	Task 60, ordnance checks
2015	ETR	Built-in hold (4 hr, 35 min)
Nov. 6 (major F – 0 day milestones)		
2349 (Nov. 5)	Spacecraft	Task VII, spacecraft radio-frequency checks
0224	ETR	Countdown initiation
0229	Delta	Task 11, launch-vehicle radio-frequency checks
0309	Spacecraft	Spacecraft standby status checks
0319	Delta	Task 12, class-A ordnance installation and hookup
0319	Spacecraft	Task IX, ordnance arm
0404	Spacecraft	Task IX, sustained operation

TABLE 2-1.—*Planned Pioneer-D Typical Detailed Task Sequence—* Continued

Date	Location	Task
0549	Delta	Tasks 13 to 15, launch-vehicle final preparation and tower removal
0549	Spacecraft	Task X, spacecraft terminal count
0649	Delta	Task 16A, liquid oxygen setup
0749	Delta	Task 17, beacon checks
0809	ETR	Built-in hold (57 min)
During hold	Delta	Task 16B, liquid oxygen fill
During hold	Delta	Task 13, second-stage pressure fill
0847	ETR	End of hold
0937	ETR	Built-in hold (5 min)
0942	ETR	End of hold
0947	ETR	Liftoff

[a] Adopted from Pioneer Specification PC–153.00. Cross checks between actual working schedules and the actual sequence of events at the Cape for Pioneer D reveal several minor changes in plans. Arabic task numbers apply to the launch vehicle; Roman numerals are assigned to spacecraft tasks. These tasks are defined in great detail in Goddard and Ames documents. Pioneer Specification PC–153.00 covered only flight operations; preparation of the spacecraft, launch vehicle, and other hardware are typified in Figure 2–7.

TABLE 2-2.—*Spacecraft Countdown Detailed Schedule*[a]

Time	Countdown time	Task
F – 2 day schedule		
0420 EST	T – 2430	All personnel report to Hangar AM.
0450	T – 2400	Area-17 personnel report to Spacecraft Coordinator in blockhouse.
0505	T – 2385	Area-17 personnel report to Levels 8B and 9.
0550	T – 2340	Start Task I, preparations and spacecraft and experiment checks.
1110	T – 2020	Complete Task I.
1110	T – 2020	TRW personnel not required in Task II report to blockhouse.
1110	T – 2020	Begin Task II, final pneumatics pressurization.
1310	T – 1900	Task II complete.
1310	T – 1900	Begin Task III, magnetometer service.
1430	T – 1820	Task III complete.
1430	T – 1820	Begin Task IV, ordnance installation and checks.
1530	T – 1760	Complete Task IV.
		End of F – 2 day activity, secure Level 8B.
F – 1 day schedule		
0040 EST	T – 1550	Crew arrives at blockhouse 30 min prior to start of Task V.
0110	T – 1520	Begin Task V, Red Tag removal and final preparations.

TABLE 2-2.—*Spacecraft Countdown Detailed Schedule*—Continued

Time	Countdown time	Task
0340	T − 1370	Complete Task V.
0340	T − 1370	TRW and NASA observers on stand as required.
0950	T − 1000	Task VI preparations; spacecraft crew reports to blockhouse.
0950	T − 1000	Begin Task VI, umbilical checks.
1010	T − 980	Task VI complete.
1040	T − 950	End of F − day activity, secure Level 8B.

F − 0 day schedule

Time	Countdown time	Task
1855	T − 525	Crew arrives at blockhouse 10 min prior to start of Task VII.
1905	T − 515	Begin Task VII, spacecraft systems checks.
2205	T − 335	Complete Task VII.
2205	T − 335	Begin Task VIII, spacecraft standby status preparations.
2215	T − 325	Complete Task VIII.
2215	T − 325	Begin Task IX, ordnance Connection and final secure.
2235	T − 305	Arm ordnance.
2255	T − 285	Secure.
2300	T − 280	Complete Task IX.
2300	T − 280	Spacecraft sustained operation.
0035	T − 185	Stand personnel required for Task X report to road block.
0045	T − 175	Start Task X, terminal countdown.
0045	T − 175	Tower removal preparations.
0200	T − 100	Tower removal.
0215	T − 85	RF checks (receiver 2).
0240	T − 60	RF checks (receiver 1).
0255	T − 45	Receiver 1 and 2 final frequency report.
0305	T − 35	Built-in hold (60 min).
0405	T − 35	Begin terminal count.
0430	T − 10	Spacecraft to internal power.
0435	T − 5	Built-in hold (5 min).
0440	T − 5	Resume count.
0442	T − 3	Spacecraft go/no-go report.
0445	T − 0	Liftoff.

[a] As issued to spacecraft launch team at Cape Kennedy for the launch of Pioneer D. Delta, DSN, and ETR events not shown, although they participate in some tests, such as the operational readiness test.

THE ACTUAL PRELAUNCH PHASES AND HOW THEY COMPARED

Each of the five prelaunch phases had its own inventory of anecdotes and special circumstances that made it slightly different from the others. Of course, the spacecraft, the Delta, the DSN, and the ETR all evolved between Pioneer flights so that the ingredients were somewhat different for each launch. A brief narrative for each launch follows with emphasis on events not appearing on the planners' charts in tables 2-1

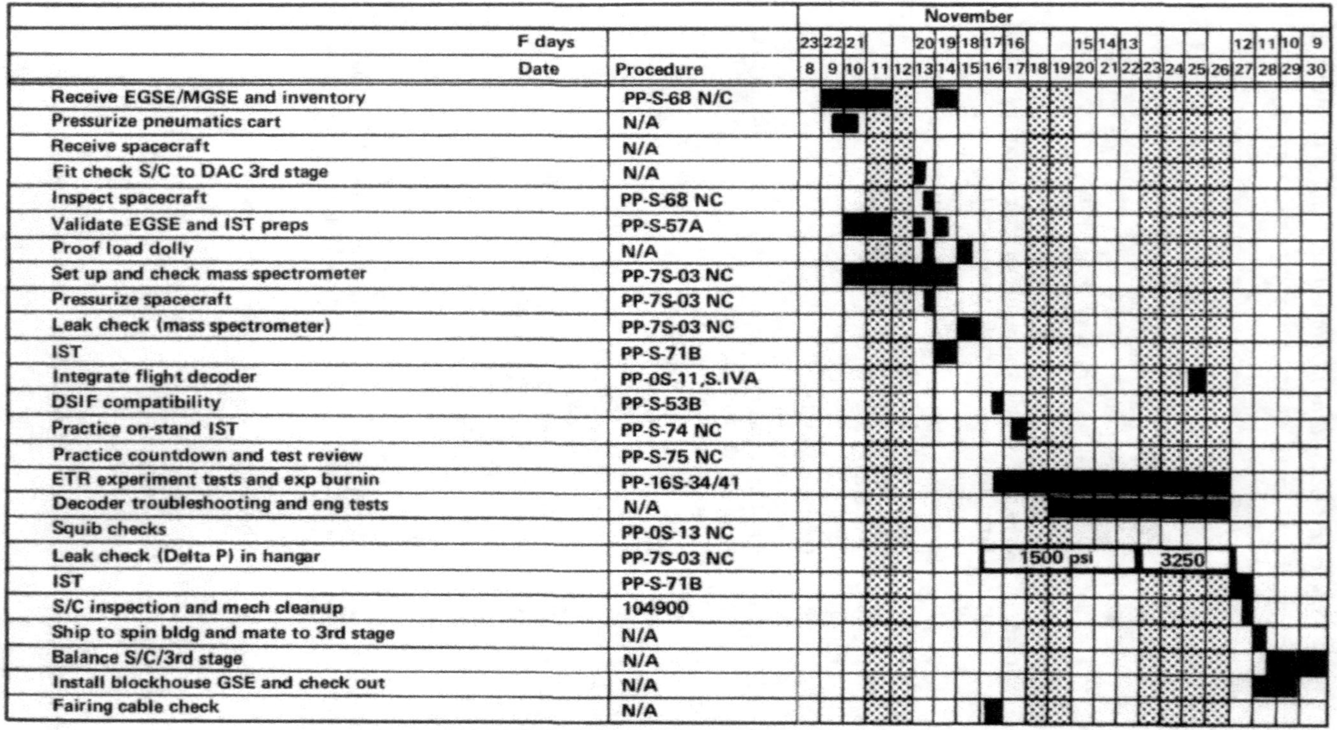

THE INTERPLANETARY PIONEERS

	F days		November																					
	Date	Procedure	23	22	21	20	19	18	17	16	15	14	13	12	11	10	9							
			8	9	10	11	12	13	14	15	16	17	18	19	20	21	22	23	24	25	26	27	28	29	30
Receive EGSE/MGSE and inventory		PP-S-68 N/C																							
Pressurize pneumatics cart		N/A																							
Receive spacecraft		N/A																							
Fit check S/C to DAC 3rd stage		N/A																							
Inspect spacecraft		PP-S-68 NC																							
Validate EGSE and IST preps		PP-S-57A																							
Proof load dolly		N/A																							
Set up and check mass spectrometer		PP-7S-03 NC																							
Pressurize spacecraft		PP-7S-03 NC																							
Leak check (mass spectrometer)		PP-7S-03 NC																							
IST		PP-S-71B																							
Integrate flight decoder		PP-0S-11,S.IVA																							
DSIF compatibility		PP-S-53B																							
Practice on-stand IST		PP-S-74 NC																							
Practice countdown and test review		PP-S-75 NC																							
ETR experiment tests and exp burnin		PP-16S-34/41																							
Decoder troubleshooting and eng tests		N/A																							
Squib checks		PP-0S-13 NC																							
Leak check (Delta P) in hangar		PP-7S-03 NC															1500 psi			3250					
IST		PP-S-71B																							
S/C inspection and mech cleanup		104900																							
Ship to spin bldg and mate to 3rd stage		N/A																							
Balance S/C/3rd stage		N/A																							
Install blockhouse GSE and check out		N/A																							
Fairing cable check		N/A																							

FIGURE 2.-7.—Pioneer-C prelaunch activities. (a) November.

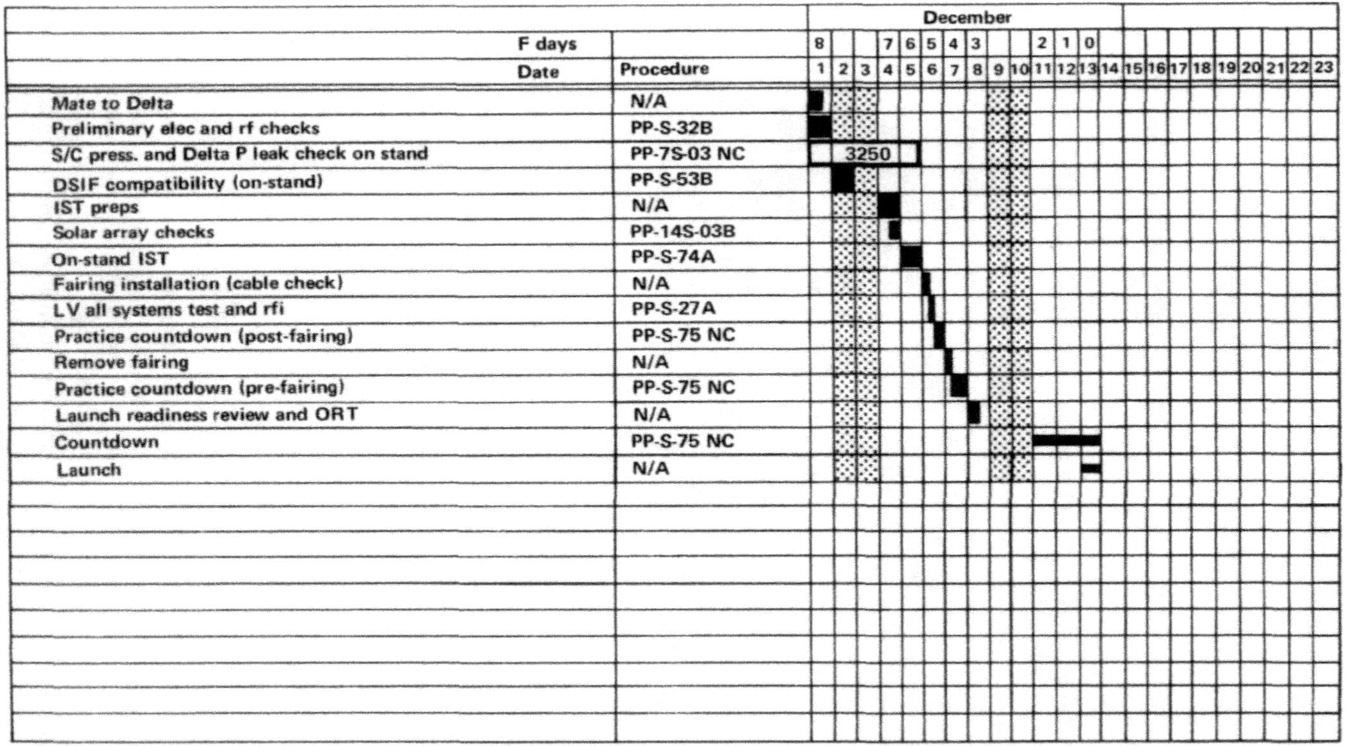

PRELAUNCH ACTIVITIES 19

F days		8	7	6	5	4	3	December 2	1	0													
Date	Procedure	1 2	3 4	5 6	7 8	9 10	11 12	13 14	15 16	17 18	19 20	21 22	23										
Mate to Delta	N/A																						
Preliminary elec and rf checks	PP-S-32B																						
S/C press. and Delta P leak check on stand	PP-7S-03 NC	3250																					
DSIF compatibility (on-stand)	PP-S-53B																						
IST preps	N/A																						
Solar array checks	PP-14S-03B																						
On-stand IST	PP-S-74A																						
Fairing installation (cable check)	N/A																						
LV all systems test and rfi	PP-S-27A																						
Practice countdown (post-fairing)	PP-S-75 NC																						
Remove fairing	N/A																						
Practice countdown (pre-fairing)	PP-S-75 NC																						
Launch readiness review and ORT	N/A																						
Countdown	PP-S-75 NC																						
Launch	N/A																						

FIGURE 2–7.—Pioneer-C prelaunch activities—Concluded. (b) December.

and 2-2. A typical schedule of actual prelaunch events is presented in figure 2-7.

Pioneer-A Prelaunch Narrative

Both the prototype and flight models were sent to the Cape. The prototype arrived October 1, 1965, and was used for practicing prelaunch operations. When the prototype model and an inert Delta third stage were mated to the launch vehicle (Delta 35) on November 29, it was discovered that the umbilical wiring was improperly connected. Modifications were made to correct this.

The Pioneer-A flight model was delivered for mating with the Delta third stage on December 5. During preliminary alignment checkout, a Total Indicator Runout of 0.25 in. was noted, indicating a physical mismatch. The attach fitting was shaved down to bring the alignment within tolerance. Tests and checkouts proceeded normally through F—1 day ("normally" meaning only minor, easily corrected problems).

December 15, F—0 day, was relatively calm with visibility of only 0.125 to 2 miles. Countdown commenced 30 min early at 1630. Everything went smoothly until T—90 min when the second-stage umbilical plug was inadvertently pulled, causing loss of power to the Delta second stage and the spacecraft itself. No one was certain what would happen if the plug were reinserted, and it was considered possible that some unforeseen signal could cause serious damage by firing some of the ordnance. The spacecraft and the Delta were revalidated. The built-in 60-min hold and ultimately the launch window had to be extended while further checks were made. The terminal count resumed at 0145, December 16, at T—35 min.

At T—2 min an abnormality in the radio guidance equipment caused another hold. The situation seemed to correct itself, and the count was recycled to T—8 min. Liftoff occurred at 0231:20 **EST**, December 16, 1965 (fig. 2-8).

Pioneer-B Prelaunch Narrative

The prelaunch operations for Pioneer B were comparatively uneventful. The flight spacecraft arrived at Building AM on July 17, 1966. On August 9, it was discovered that, when the Chicago cosmic-ray experiment warmed up, a connection opened, partially disabling the experiment. As a result, the experiment indicated a non-existent low radiation level at all times. The experiment flew in this condition.

On F—2 day, August 15, a receiver lockup problem was encountered on the two S-band uplinks. Ultimately, the trouble was traced to an antenna on Building AM that was not pointed toward the launch pad.

F—0 day, August 17, had superb weather, with 5-knot winds and

FIGURE 2-8.—The launch of Pioneer A on Delta 35.

a visibility of 10 miles. The countdown proceeded normally to T−3 min, when a hold was called due to the loss of communications downrange on the ETR. Communications were restored after 2 min, and liftoff occurred at 1020:17 EST.

Pioneer-C Prelaunch Narrative

Pioneer C was the first of the Block-II spacecraft. In addition, this flight was the first to carry a TETR mounted in the Delta second stage. The Pioneer-C flight model was received at Building AM on November 11, 1967. The IST of November 15 identified a faulty decoder which was replaced. On November 22, the Ames plasma probe was removed to correct a wiring error.

FIGURE 2-9.—Pioneer E arriving by air-freight from the West Coast.

F — 2 day, December 11, was plagued by bad weather which forced personnel to clear the pad twice. At 1520, the electrical power was lost for 25 min, causing some concern because the spacecraft air conditioning was also lost. On F — 1 day, the fairing had to be removed to repair wiring to the third-stage velocity meter. Terminal count began at 0543, December 13, and Pioneer C was launched successfully at 0908:00 on December 13, 1967.

Pioneer-D Prelaunch Narrative

This spacecraft was the first to incorporate the convolutional coder experiment and the Ames magnetometer. Pioneer D arrived at Building AM on October 6, 1968. The beginning of the countdown was delayed for 2 days while tests and adjustments were made to the second-stage programmer. As soon as the programmer was accepted for flight, F—2 day activities commenced. The countdown proceeded smoothly to 0900 EST when anomalies appeared in the experimental data and experimental performance. Holds were called to investigate these problems which were found to be due to radio and electrical interference from the launch vehicle. No troubles were encountered during F—1 day countdown activities. At 1850 EST, November 7, 1968, F—0 day checks began. Spacecraft power was turned on at 1920. Spacecraft systems checks (Task VII) ran ahead of schedule, and a 20-min hold was called at 2015 to give the spacecraft receiver additional time to warm up. The terminal count began at 0050, November 8. Following a hold of 9.5 min due to high sheer winds aloft, the Delta lifted off at 0446:29.

Pioneer-E Prelaunch Narrative

On July 18, 1969, the Pioneer-E spacecraft was received at Cape Kennedy (fig. 2–9). There were no unusual prelaunch events. A study of the launch vehicle test summary indicates a normal sequence of prelaunch events although a number of minor problems arose—as they usually do—and were corrected. Nothing in the prelaunch tests and checkout presaged the failure of the launch vehicle after liftoff.

Spacecraft and radio-frequency checks, Task VII, began at 0835 EDT on F—0 day, August 27, 1969. Except for a thunderstorm that temporarily delayed work, weather was excellent, with a visibility of 8 miles and light winds. The terminal countdown was uneventful. Liftoff was at 1759:00 EDT, August 27, 1969.

CHAPTER 3

Launch to DSS Acquisition

THE PHASE OF PIONEER OPERATIONS stretching from launch to DSS acquisition lasted less than 1 hr, but it was the only time when all four Pioneer systems operated together. Even this observation is a forced one because the spacecraft and scientific instrument systems were essentially passive during powered flight and the coast phase. Only housekeeping data were telemetered, and all scientific instruments were off. The spacecraft came to life only when the Travelling Wave Tubes (TWTs) were switched on, the booms were deployed, and the Type-I orientation maneuver was initiated. By this time, the spacecraft had been spun up and had separated from the Delta third stage. The ground-based DSN was involved in a configuration called the Near-Earth Phase Network which, during this phase, incorporated some facilities from the Air Force Eastern Test Range and the NASA Manned Space Flight Network (MSFN). Figure 3-1 illustrates the chronology and terminology involved in this phase of the mission.

It is best to view Pioneer operations from several vantage points so that the operations of all four systems can be appreciated. First, the sequence of events is portrayed schematically in figure 3-2. The nominal time frame for one of the launches is added to the picture in table 3-1. Of course, the timing of the critical events varied from mission to mission because the burn and coast times changed with each launch, and the Delta rocket was upgraded during the program. The nominal time frame, with its critical events, provides a yardstick against which to measure the success of this portion of the launch. Altitude and distance downrange are also important diagnostic parameters. However, the DSN stations waiting downrange tend to see the picture as having an additional spatial dimension (fig. 3-3). This is not surprising because the tracking of spacecraft in orbit or far out in space is essentially four-dimensional (including time), whereas range tracking, as on the ETR, was more nearly three-dimensional in character; that is, altitude, range, and time. In fact, one of the critical displays at the Cape is an impact point predictor—a two-dimensional display. "Handover" from the Near-Earth Phase Network to the DSN was facilitated by "predicts" sent ahead to DSS-51 (Johannesburg) to tell its acquisition-aid antennas where to look in the western sky. Once the acquisition aid had the spacecraft, the 85-ft. parabolic antenna was slewed to it.

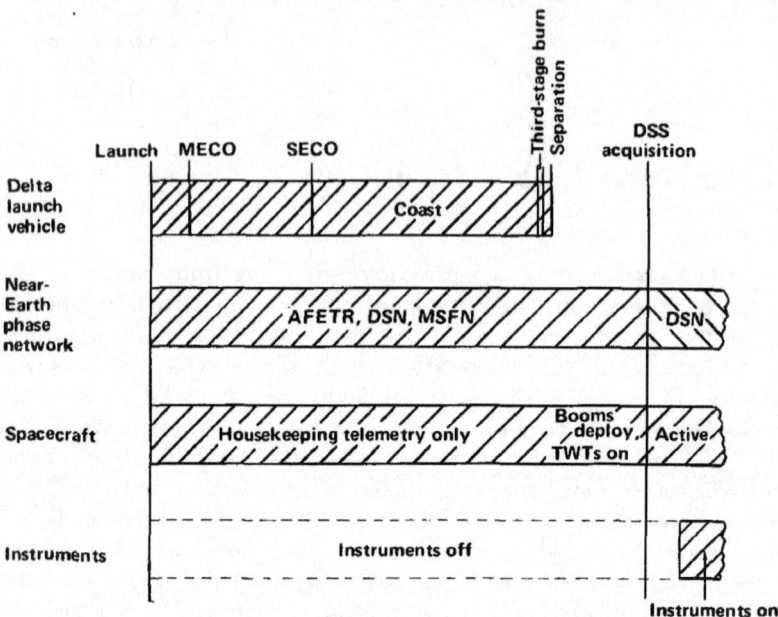

FIGURE 3-1.—Status of the four Pioneer systems from launch through DSS acquisition.

PERFORMANCE OF THE DELTA LAUNCH VEHICLE

The Delta launch vehicle performed superbly during the first four Pioneer launches. The fifth mission, Pioneer E, had to be aborted by the Range Safety Officer when the vehicle began to stray off course. It would be repetitious to narrate each launch in detail. Instead, tables 3–2 and 3–3 summarize Delta "mark events" and stage performance, respectively. A series of figures (figs. 3–4 to 3–7) portrays the Delta overall metric performance for the four successful missions. Of course, none of the flights was flawless, but the deviations were all minor compared to the failure on the final flight. These perturbations are summarized below along with a description of the loss of Pioneer E.

Pioneer-6 Launch Vehicle Performance

All mission objectives were achieved and vehicle flight was well within the "three-sigma" limits.[5] All liquid engines and solid motors performed satisfactorily, including the spinup rockets. Second-stage thrust misalignment in pitch was excessive, however, and alignment procedures were to be modified on future Delta flights. Another minor deviation

---

[5] Theoretically, the "three-sigma" limits encompass 99.86 percent of all observations.

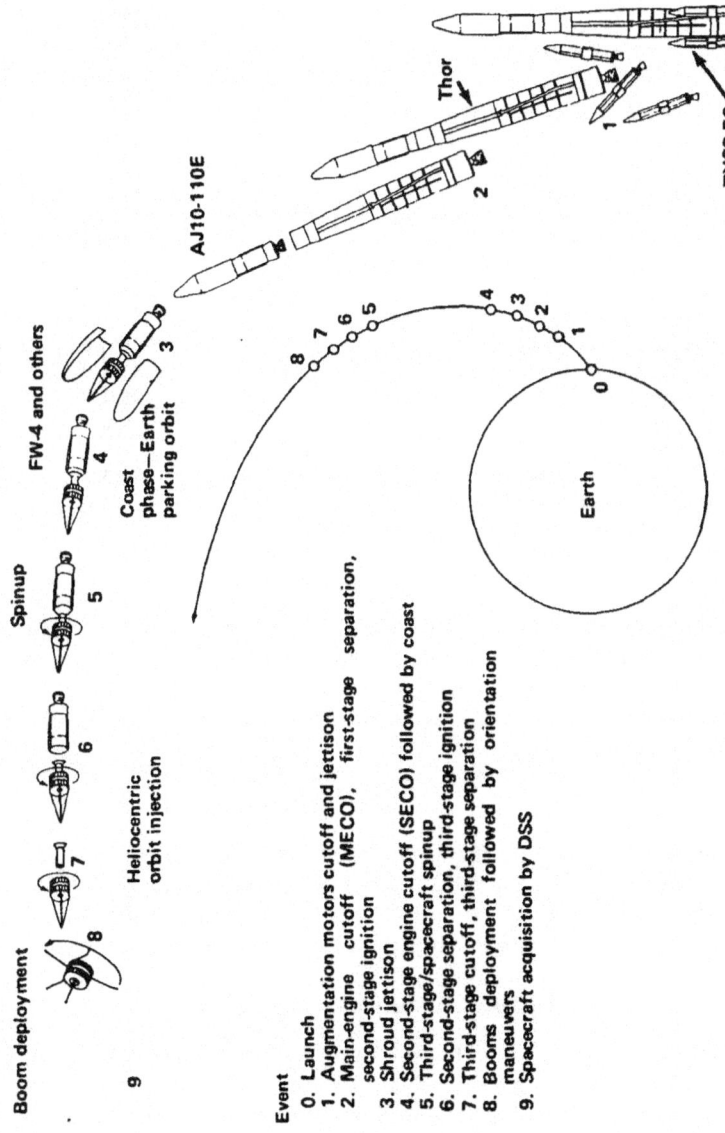

FIGURE 3-2.—A typical Pioneer launch sequence.

TABLE 3-1.—*Pioneer-6 Nominal "Mark Events," Launch to DSS Acquisition*

Mark event	Nominal time from launch, sec
Liftoff	0
Solid motor burnout (thrust augmentation)	43.0
Grand Bahama rise (ETR station)	56.0
Solid motors jettisoned	70.0
Main engine cutoff	149.2
Second-stage ignition	153.2
First-stage jettisoned	153.4
Fairing jettisoned	179.2
Second-stage engine cutoff	551.1
Control transferred from SFOF to Johannesburg	1200.0
Third-stage spinup	1486.2
Second-stage jettisoned	1488.2
Third-stage ignition	1501.2
Third-stage burnout	1523.7
Spacecraft separation	1583
Type-I orientation begins automatically	1583
Booms and Stanford antenna deployed	1584
TWT 1 turned on and switched to low-gain antenna	1584
Johannesburg rise (DSS acquisition can occur after this time)	1688

from normal performance was caused by the asymmetric ejection of the nozzle closure ring on the second-stage engine. As a result, a redesign of the nozzle closure diaphragm was recommended. These minor departures from perfect performance are common in any operation involving sophisticated machines.

Pioneer-7 Launch Vehicle Performance

The launch vehicle performed even better on this flight than it did on Pioneer 6. The only failure of even minor significance concerned the third-stage air-inlet-adapter primary-loop lanyard, which failed during liftoff. The adapter was successfully released by a secondary lanyard.

Pioneer-8 Launch Vehicle Performance

Pioneer 8 and the TETR–1 satellite were launched successfully. All flight parameters were well within the three-sigma limits. Unexplained moments in the pitch and yaw planes were noticed just after second-stage engine cutoff, but they did not affect the mission. Some circuit anomalies occurred after second-stage separation. These were caused when the second-stage engine burned some wire insulation on the jettisoned first stage.

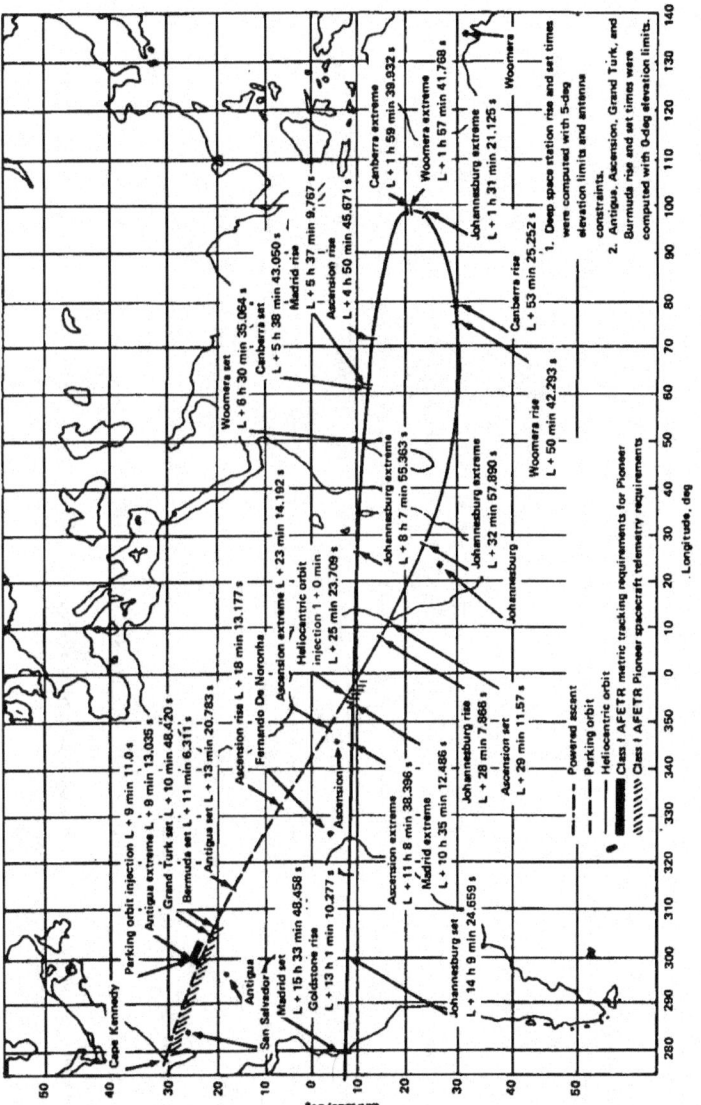

FIGURE 3-3.—Pioneer-6 ground track showing tracking station coverage.

TABLE 3–2.—*Summary of Critical Nominal and Actual Launch Events*

Event	Pioneer 6[a] Nominal	Pioneer 6[a] Actual	Pioneer 7[b] Nominal	Pioneer 7[b] Actual	Pioneer 8[c] Nominal	Pioneer 8[c] Actual	Pioneer 9[d] Nominal	Pioneer 9[d] Actual	Pioneer E[e] Nominal	Pioneer E[e] Actual
Liftoff (sec)	0	0	0	0	0	0	0	0	0	0
Solid motor burnout	43.00	42.23	42.65	41.60	41.90	42.25	38.19	39.4	70.00	69.8
Solid motors jettisoned	70.00	69.97	70.00	70.79	70.00	70.3	70.00	69.94[f]	220.0	215.4
Main engine cutoff	149.21	148.01	149.52	148.10	149.75	149.45	150.53	151.35	224	224
Second-stage ignition	153.21	152.03	153.52	152.13	153.75	153.50	154.53	155.42	229	229.1
Fairing jettisoned	—	178.84	175.52	174.74	185.75	186.16	169.53	171.01		
Second-stage engine cutoff command	551.13	531.44	529.32	527.89	530.93	536.73	534.35	539.01		
Third-stage spinup	1486.21	1485.03	1575.52	1474.12	1842.75	1842.42	1201.53	1202.40		
Second-stage jettisoned	1488.21	1487.03	1477.52	1476.12	1844.75	1844.43	1203.53	1204.37		
Third-stage ignition	1501.21	1496.94	1490.52	1488.98	1857.75	1857.98	1216.53	1218.54	—	—
Third-stage burnout	1523.71	1520.74	1521.34	1519.80	1888.55	1888.78	1247.33	1249.34	—	—
TETR separation	—	—	—	—	1904.75	1904.25	1263.53	1264.4	—	—
Actual launch date	Dec. 16, 1965		Aug. 17, 1966		Dec. 13, 1967		Nov. 8, 1968		Aug. 27, 1969	
Actual launch time	0231:20 EST		1020:17 EST		0908:00 EST		0446:29 EST		1759:00 EST	

[a] McDonnell-Douglas Corp.: Flight Report for Improved Delta Vehicle S/N 460/20202/20202, Delta Program—Mission No. 35, Spacecraft: Pioneer A, Rept. SM-48998, 1966, pp. 44–46.
[b] McDonnell-Douglas Corp.: Flight Report for Delta Vehicle S/N 462/20204/20207, Delta Program—Mission No. 40, Spacecraft: Pioneer B, Rept. DAC-58324, 1966, pp. 30–31.
[c] McDonnell-Douglas Corp.: Flight Report for Delta Vehicle S/N 20216, Delta Program—Mission No. 55, Spacecraft: Pioneer C, Rept. DAC-58726, 1968, pp. 45–47.
[d] McDonnell-Douglas Corp.: Flight Report for Delta Vehicle S/N 20222, Delta Program—Mission No. 60, Spacecraft: Pioneer D, Rept. DAC-61695, 1969, pp. 35–37.
[e] First-stage hydraulic pressure lost at 214 sec. Destruction by Range Safety Officer at 483.9 sec.
[f] For one motor; the other two augmentation motors were jettisoned at 69.992 sec.

TABLE 3-3.—Summary of Delta Performance Factors[a]

	Pioneer 6		Pioneer 7		Pioneer 8		Pioneer 9	
	Nominal	Actual	Nominal	Actual	Nominal	Actual	Nominal	Actual
Overall weight (lb)	149,977	149,904	150,084	149,560	149,537	149,468	151,765	151,875
First-stage liquid-engine thrust (lb)	186,889	188,567	186,525	187,803	186,967	187,502	185,888	185,432
First-stage liquid-engine specific impulse (sec)	273.12	273.13	279.39	279.58	281.16	280.98	281.72	281.72
MECO[b] inertial velocity (ft/sec)	14,141	14,153	12,941	13,020	14,049	14,093	14,331	14,242
MECO inertial flight path elevation angle (deg)	16.9	16.9	17.23	17.02	20.34	20.31	19.38	19.32
MECO inertial flight path azimuth angle (deg)	100.9	101.0	107.31	107.41	107.22	107.15	107.31	107.24
MECO range (n. mi.)	91.1	90.8	92.7	92.7	87.9	87.9	92.4	92.4
MECO altitude (n. mi.)	48.3	48.1	51.0	51.3	54.9	54.4	55.9	55.4
Second-stage thrust (lb)	7369.0	7384.8	7553.5	7485.2	7572.6	7502.2	7527.3	7371.3
Second-stage specific impulse (sec)	268.9	269.1	271.84	274.41	272.3	279.4	275.2	267.8
SECO inertial velocity (ft/sec)	26,348	26,279	26,056	26,065	25,342	25,216	25,677	25,715
SECO inertial flight path elevation angle (deg)	0.033	0.087	−0.61	−0.86	0.97	0.75	0.10	−0.06
SECO inertial flight path azimuth angle (deg)	110.6	111.1	115.96	116.12	115.69	115.83	115.91	115.73
SECO range (n. mi.)	1185	1119	1109.1	1111.1	1063.1	1079.8	1099.5	1108.0
SECO altitude (n. mi.)	149.7	149.2	149.9	145.8	205.9	205.5	202.9	202.5
Third-stage burn time (sec)	22.5	23.8	30.82	30.82	30.80	30.80	30.80	30.80
Third-stage impulse (lb-sec)	139,761	140,496	172,160	171,896	172,696	172,418	172,481	172,354
TSB inertial velocity (ft/sec)	35,542	35,593	35,869	35,898	35,185	35,383	36,225	36,208
TSB inertial flight path elevation angle (deg)	1.6	1.7	2.44	2.01	0.00	−0.55	2.00	2.22
TSB inertial flight path azimuth angle (deg)	119.2	119.3	129.51	129.81	114.60	114.35	124.33	124.53
TSB range (n. mi.)	4913.8	4898.5	4952.0	4980.1	5984.8	6025.3	3794.6	3794.5
TSB altitude (n. mi.)	301.7	304.6	230.2	204.0	314.7	262.3	251.5	251.8

[a] Same references as those cited in table 3-2.
[b] MECO = Main Engine Cutoff; SECO = Second-stage Engine Cutoff; TSB = Third Stage Burnout.

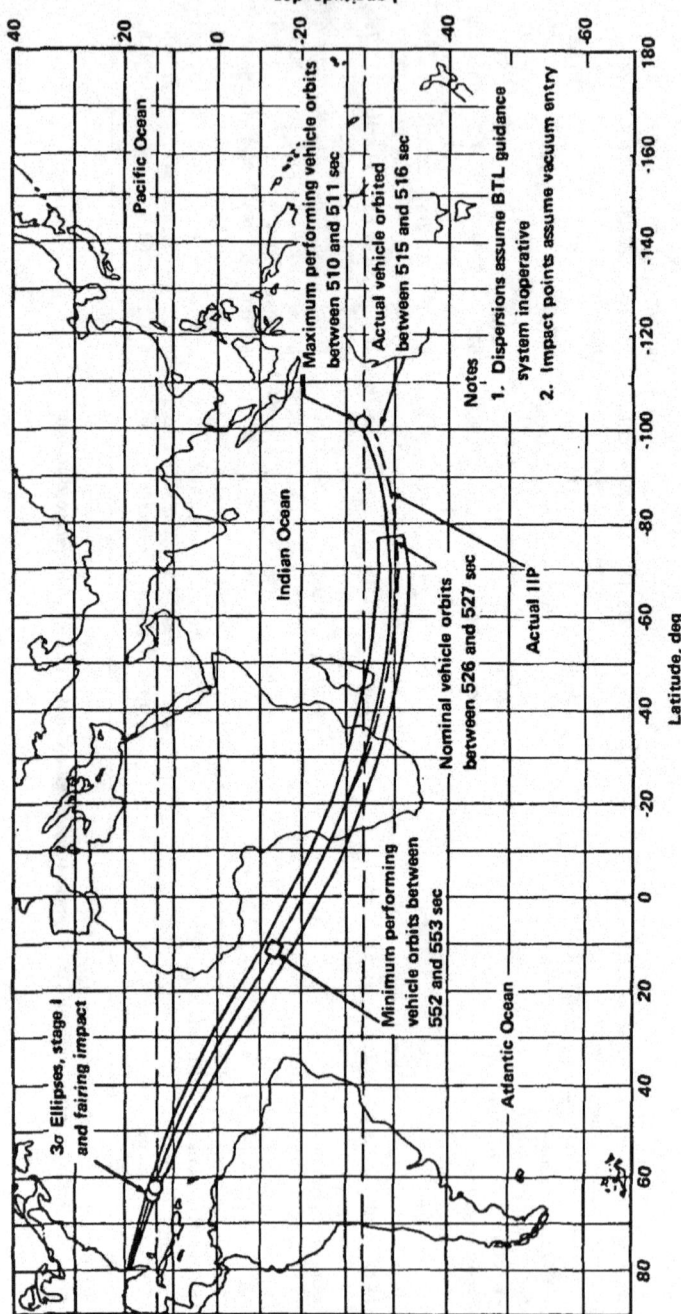

FIGURE 3-4.—The Pioneer-6 trajectory, showing the 3σ limits.

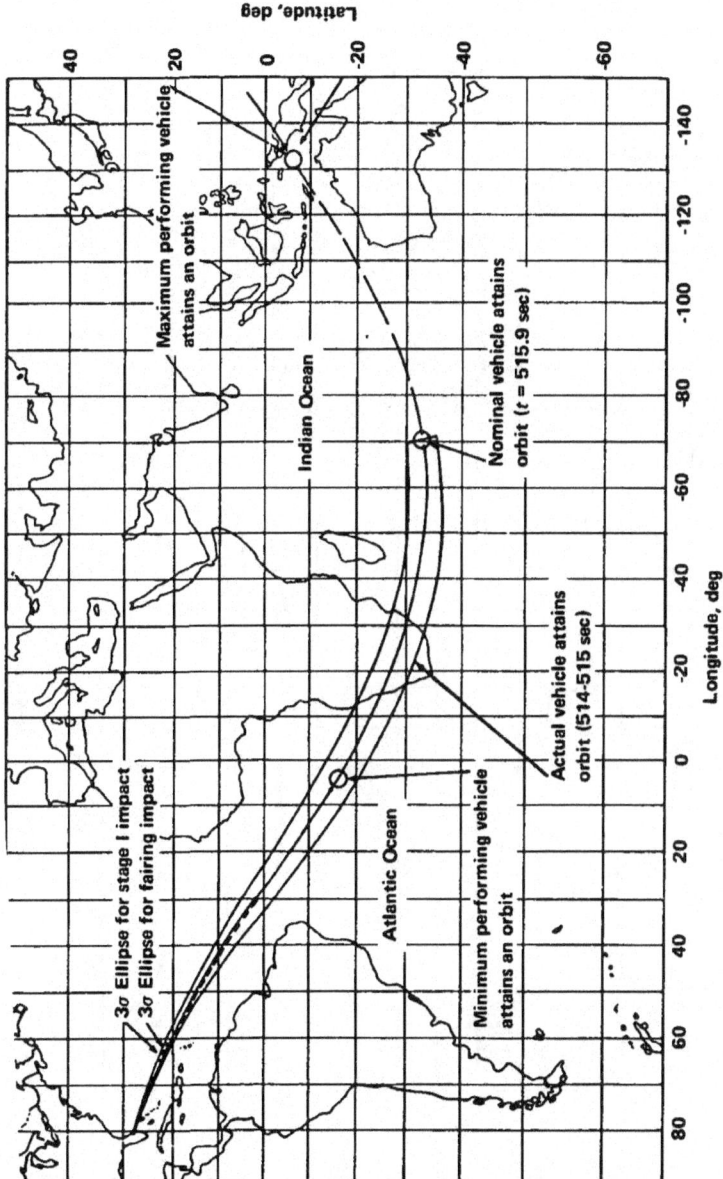

FIGURE 3-5.—The Pioneer-7 trajectory, showing the 3σ limits.

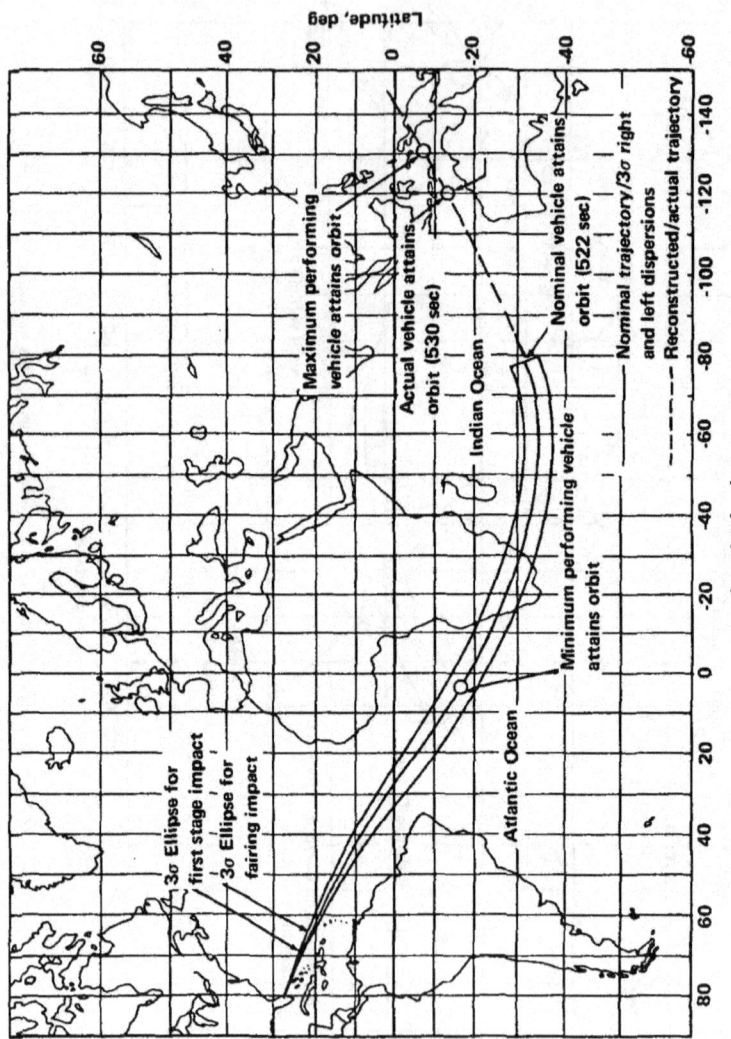

FIGURE 3-6.—The Pioneer-8 trajectory, showing the 3σ limits.

LAUNCH TO DSS ACQUISITION 35

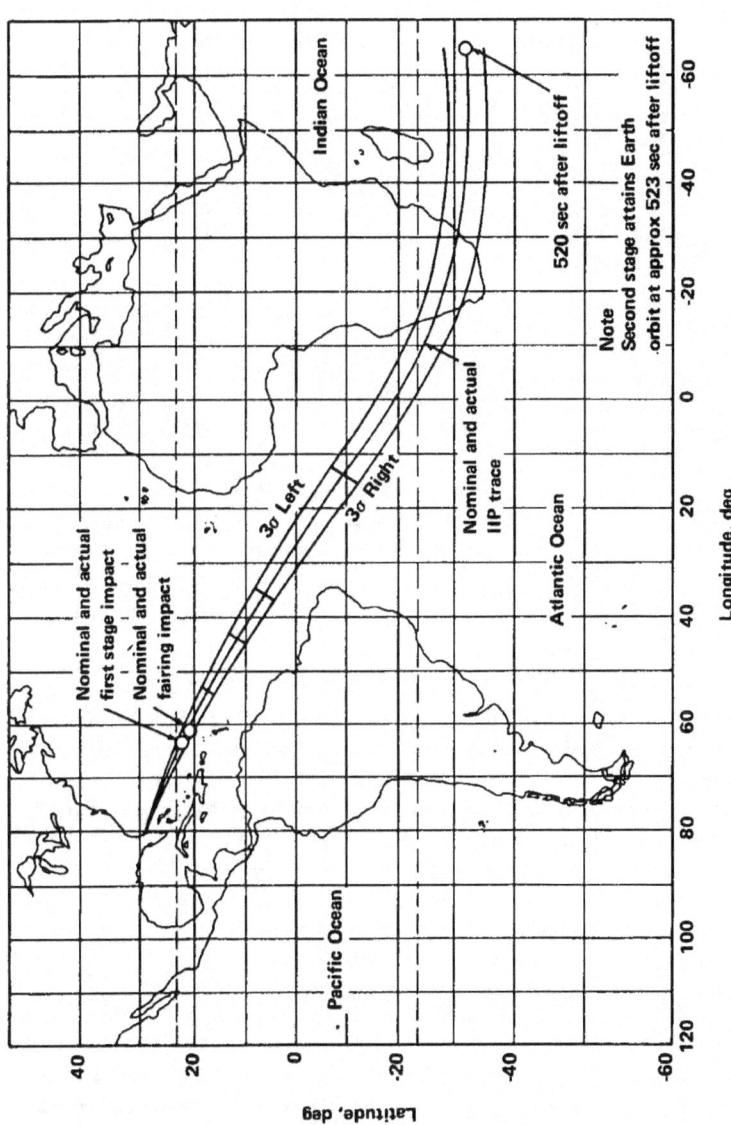

FIGURE 3-7.—The Pioneer-9 trajectory, showing the 3σ limits.

Pioneer-9 Launch Vehicle Performance

This was essentially a perfect launch from the Delta viewpoint.

Pioneer-E Launch Vehicle Performance

Ignition and liftoff were normal. The three solid, thrust-augmentation engines operated properly and were jettisoned simultaneously and at the proper moment. At 63 sec into the flight, the first-stage hydraulic pressure decreased from 3150 to 3000 psia. Then the pressure began fluctuating. At 213 sec, the pressure dropped to zero, and all first-stage control was lost. Telemetered propulsion parameters began to indicate violent vehicle maneuvers.

With the first-stage hydraulic pressure lost, the main engine pitched down, yawed left, and rolled counterclockwise. As a result, the second-stage gyros were driven out of their limits during second-stage ignition and separation. The ignition and initial performance of the second-stage engine were approximately normal, but the damage was done, and the second stage was far off course.

After main engine cutoff, the predicted impact point on the plotboard began to move at a 45° angle to the right and downrange. The Range Safety Officer sent "arming" commands to downrange stations to ensure they had the capability for vehicle "destruct." At 483.9 sec, the deviation from the planned course was too great, and the vehicle was destroyed.

TRACKING AND DATA ACQUISITION

As a Pioneer spacecraft and its launch vehicle rose from the launch pad at Cape Kennedy, they were tracked downrange by a variety of radio and optical tracking devices. Until the spacecraft was "handed over" to the Johannesburg DSS station, the pooled radars, optical trackers, guidance equipment, and telemetry receivers of the Air Force ETR and some stations of NASA's DSN and MSFN were crucial to mission success. We examine now how these resources were put together and how they functioned during this phase of each Pioneer mission.

Tracking and Data Acquisition Requirements and Station Configurations

Tracking and telemetry data are needed to assess the performance of both the launch vehicle and the spacecraft and also for ensuring range safety. All of the "mark events" and launch vehicle positions, velocities, and headings described in the preceding section are obtained through tracking and analysis of telemetry data.

The facilities assigned to each of the Pioneer missions from launch through DSS acquisition are listed in table 3–4. The ETR was the pri-

LAUNCH TO DSS ACQUISITION 37

TABLE 3-4.—*Tracking and Data Acquisition Support Stations Through DSS Acquisition*

Range/network	Station	Used during Pioneer flights				
		6	7	8	9	E
AFETR	1. Cape Kennedy and Patrick AFB	x	x	x	x	x
	3. Grand Bahama I.	x	x	x	x	x
	7. Grand Turk I.	x	x	x	x	x
	9.1 Antigua I.	x	x	x	x	x
	12 Ascension I.	x	x	x	x	x
	13 Pretoria, S.A.	x	x	x	x	
	Twin Falls (ship)	x	x	x		
	Coastal Crusader (ship)	x	x			
	Sword Knot (Ship)	x				
MSFN	Merritt I.					x
	Bermuda	x	x	x	x	x
	Grand Bahama					x
	Antigua					x
	Ascension I.	x	x	x	x	x
	Tananarive, Malagasy Rep.	x	x	x	x	x
	Vanguard (ship)					
DSN	DSS-71, Cape Kennedy	x	x	x	x	
	DSS-72, Ascension I.	x	x	x		
	DSS-51, Johannesburg, S.A.	cx	x	x	x	ax
	DSS-41, Woomera, Australia [b]			x		

[a] Scheduled, but not actually used due to abort.
[b] The "primary" DSN acquisition station for Pioneer 8.
[c] Commanded partial Type-II orientation. This maneuver was commanded from Goldstone on Pioneer 9.

mary agency responsible for providing metric (tracking) data during this phase of operations. The MSFN stations, listed in table 3-4, provided redundant radar support. Metric requirements were met by tracking the C-band beacon aboard the Delta and the S-band telemetry signal from the spacecraft. From liftoff to 5000-ft altitude, ETR optical equipment provided additional metric data.

During this "powered" phase of flight, telemetry came primarily from the first and second stages of the launch vehicle. These were PDM/FM/FM links operating at 228.2 and 234.0 MHz, respectively. ETR stations 1, 3, 9.1, 12, 13, and the Range Instrumentation Ships (RIS) acquired this telemetry.

The ETR Real Time Computer Facility (RTCF) at the Cape, which used CDC 3600 and 3100 computers during the Pioneer flights, processed the metric data flowing back from downrange sites and converted them into "predicts" for stations which had not yet acquired the spacecraft and/or launch vehicle stages.

During the Pioneer flights, Building AO (fig. 3-8) contained joint JPL–AFETR facilities that were vital to mission analysis and control. Briefly, these facilities were:

(1) A joint operations center consisting of status displays, a timing system, and consoles. During the early portion of the Pioneer mission, control was transferred to the SFOF.

(2) A joint communication center, which provided the local terminals and interfaces for voice, teletype, and data circuits to and from Cape facilities, the SFOF, and the ETR.

Flight Operations—Tracking and Data Acquisition

The scenarios of tracking and data acquisition activities vary slightly for each launch. Short narratives rather than tables seem in order here.

Pioneer 6.—Liftoff occurred at 0731:20 GMT, December 16, 1965. Grand Bahama rise was at 0732:15, but its receiver was in and out of lock until 0737:33. At 0751, control was transferred from the SFOF to DSS–51, Johannesburg, for the planned partial Type-II orientation. The Johannesburg acquisition aid antenna acquired the spacecraft at 0759. Initial telemetry indicated that the automatic Type-I orientation was underway. At 0804:14, the spacecraft signaled that the Type-I orientation was complete. At 0807, the Johannesburg receiver was transferred from the acquisition aid antenna to the 85-ft dish.

There were some minor problems with telemetry coverage during this flight. Stations did not acquire and lock onto the spacecraft signal as early as predicted; however, the signal was tracked longer than anticipated (fig. 3-9). This anomalous performance was scheduled for later investigation. The first and only indication of third-stage ignition

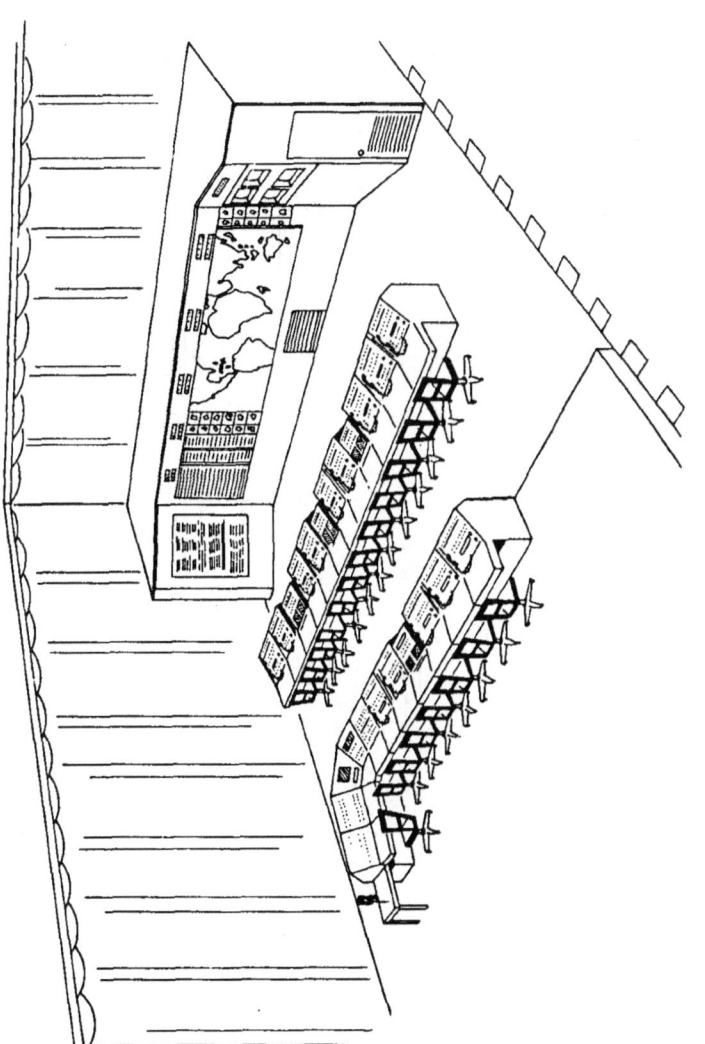

FIGURE 3-8.—Sketch of the JPL-AFETR Operations Center in Building AO at Cape Kennedy.

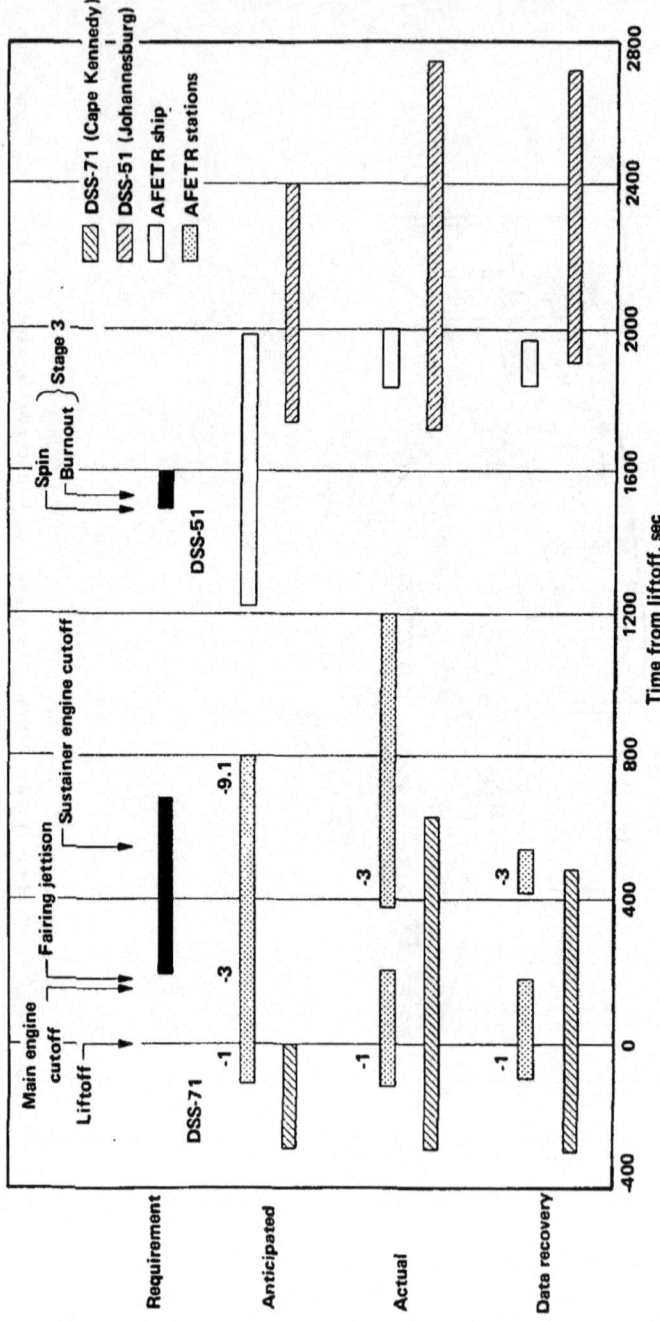

FIGURE 3-9.—Pioneer-6 telemetry coverage from launch through DSS acquisition.

in near real time was the dropout of vhf telemetry as the third-stage plume engulfed the second stage. RF propagation problems precluded receipt of real-time data from Ascension Island. When a report was finally received, it indicated only a 50-percent burn time. All other tracking and telemetry data, however, indicated a normal third-stage burn. Later, it was found that someone had read the wrong scale on the Doppler plot at Ascension. Finally, the *Coastal Crusader* obtained no Doppler data from either the 136-MHz Delta beacon or the spacecraft S-band carrier due to equipment problems. The parking orbit parameters computed from downrange tracking data are tabulated in table 3–5.

Pioneer 7.—This mission's launch window opened at 1518; liftoff was at 1520:17, August 17, 1966. Grand Bahama rose at 1521. Nominal Johannesburg rise was 1547, but acceptable two-way Doppler was not achieved until 1558:24. Telemetry indicated that the Type-I orientation had been completed and had required 28 gas pulses.

Near-Earth tracking was generally excellent, except for some dropouts as indicated in figure 20a. Due to an operator error aboard the *Sword Knot*, the spacecraft separation mark event was not recorded. This error was attributed to the fact that the *Sword Knot* did not receive mission instructions until F−1 day. Early in the launch phase, the spacecraft S-band telemetry signal was some 20 dB higher than expected.

Pioneer 8.—Under an overcast sky, liftoff occurred at 1408:00 GMT, December 13, 1967. The flight appeared to be on time until 480 sec into the flight, when the African Destruct Line was crossed 6 to 8 sec later than predicted. At DSS–51 (Johannesburg), only one-way downlink contact was made to check the spacecraft status, which was normal. Because of the short pass and excessive tracking rates at Johannesburg caused by the location of the trajectory, DSS–41 (Woomera) was considered the primary station for first acquisition. Woomera rise and acquisition were at 1455:42. Two-way lock was established at 1510:52. Telemetry indicated that the Type-I orientation had been completed successfully. The first command to the spacecraft was sent at 1535.

A special third-stage telemetry system was flown on this mission. It provided better data on the events from third-stage spinup through spacecraft separation. The third stage proved difficult to track because it apparently began tumbling after spacecraft separation. The parking orbits given in table 3–5 were computed from Antigua data. Predicts for DSS–51 and DSS–41 were also generated from Antigua data with the addition of nominal third-stage burn parameters.

Pioneer 9.—Liftoff occurred at 0946:29, November 8, 1968. At 0948:40, DSS–71 (Cape Kennedy) and ETR station reported they had lost the

TABLE 3-5.—*Earth Parking Orbit Parameters*

Parameter	Pioneer 6		Pioneer 7		Pioneer 8		Pioneer 9	
	Nominal	Actual	Nominal	Actual	Nominal	Actual	Nominal	Actual
Eccentricity	0.07682	0.07181	0.05383	0.05479	0.0213	0.0139	0.0424	0.0424
Inclination (deg)	30.168	30.196	32.915	33.005	32.91	32.90	32.9	32.876
Period (min)	101.375	100.604	97.6	97.4			98.4	98.1
Apogee (n. mi.)	743.1	704.8	533	552	324	267	502	525
Perigee (n. mi.)	150	149.1	145	137	165	166	203	202

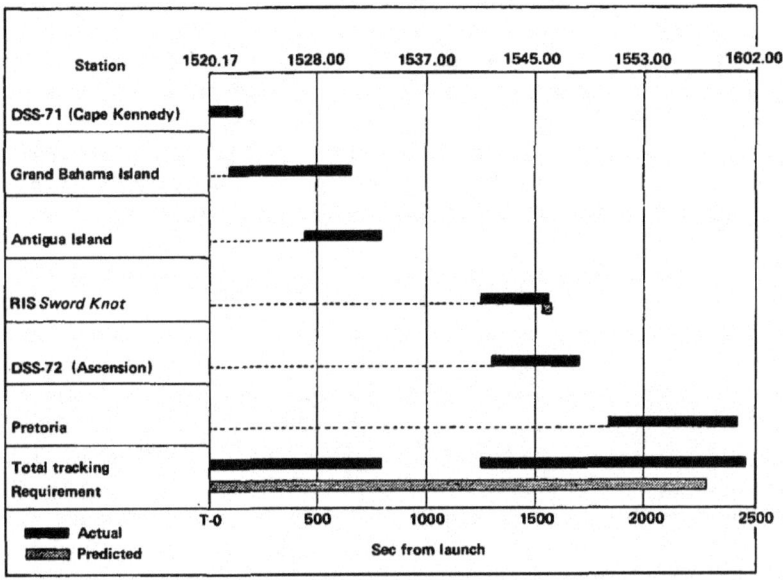

FIGURE 3-10.—Pioneer-7 tracking coverage along the ETR.

spacecraft's S-band signal. The supposition was that the vibration of the launch had thrown the spacecraft into a nonstandard position, but a subsequent investigation failed to determine the exact cause. At 1012:32, DSS-51 (Johannesburg) reported a momentary signal; this was about 10 min later than the predicted acquisition time, and the signal was 16 dB low. Two-way lock was established at 1030:18. The first command was sent at 1045:23. The spacecraft transmitter power level later rose to normal levels.

Cape Kennedy radar coverage terminated early in order to provide a phasing slot for another radar. The Antigua radar lost track for 5 sec due to an unexplained power fluctuation. Generally, though, the radar coverages exceeded expectations.

The vhf telemetry support also exceeded predictions. Grand Bahama lost 18 sec of data when the antenna slewed away from the vehicle during a switchover. The *Coastal Crusader* could not produce real-time data on two VHF links due to the failure of a recorder.

Pioneer E.—Following a brief thunder shower, liftoff took place at 2159:00, August 27, 1969. Mainland radars followed the flight through destruct, which was commanded at 482 sec, to 1136 sec. Radars at ETR stations 3 and 7 provided additional coverage from 85 to 1257 sec. Other ETR downrange stations did not acquire the vehicle. The launch vehicle, the Pioneer-E spacecraft, and the TETR-C impacted in the Atlantic at 11°30.23'N, 55°42.1'W.

During the period following main engine cutoff, the radars tracked something that veered off at a 45° angle to the right of the vehicle trajectory. The identity of this object could not be determined.

Analysis of telemetry from ETR stations, DSS-71 (Cape Kennedy), and downrange MSFN stations indicated that the spacecraft was operating normally at the time the destruct command was sent.

SPACECRAFT PERFORMANCE

The spacecraft were nearly dormant during the so-called "powered-flight" stages. About 5 min before launch, the spacecraft was put on internal power, that is, the battery. The spacecraft low-gain antenna 2 was connected to the transmitter driver rather than one of the TWTs in order to conserve battery power. Consequently, only about 40 mW of signal power was broadcast until the TWT was switched on. Housekeeping telemetry during launch was set at 64 bps—a relatively low rate—in order to increase the likelihood of obtaining good diagnostic data at a low power level if the TWT failed to turn on properly.

As soon as the spacecraft separated from the Delta third stage, the booms and Stanford antenna automatically deployed and locked into position. Power was applied to the TWT and the orientation subsystem, again automatically. The Type-I orientation maneuver then began and proceeded in the manner described in Ch. 4, Vol. II. When the low-gain antenna was switched from the transmitter driver to the TWT, the telemetry signal from the spacecraft faded for about a minute while the TWT warmed up. By the time Johannesburg rose, the spacecraft was transmitting about 7W. It was fully operational and had completed one Type-I orientation maneuver. Upon acquisition, the first commands generally sent were: (1) switch to 512 bps, and (2) repeat the Type-I orientation maneuver.

Pioneers 6 through 9 successfully went through the above sequence of events with the exception of Pioneer 9 which experienced the switching problem described earlier.

CHAPTER 4

From DSS Acquisition to the Beginning of the Cruise Phase

SEQUENCE OF EVENTS

THE PERIOD OF SEVERAL HOURS between the initial acquisition of the spacecraft by one of the DSSs and the beginning of the cruise phase encompassed several events crucial to the success of the mission:
 (1) Two types of orientation maneuvers
 (2) Experiment turn ons
 (3) The first thorough assessment of spacecraft operational condition in flight
 (4) The first passes over all participating DSSs.

Prior to DSS acquisition, the spacecraft automatically went through the Type-I orientation maneuver. This event was started by microswitches triggered when the deploying appendages locked into position. By the time the spacecraft was acquired by DSS, spacecraft power was on and the transmitter was sending telemetry. In addition, the spin axis was almost perpendicular to the Sun line by virtue of the automatic Type-I orientation maneuver.

The first command dispatched after a two-way lock had been established was usually that which changed the telemetry bit rate from Format C, 64 bps, to Format C, 512 bps. Next, a command initiating the Type-I orientation maneuver was sent to refine the alignment made automatically prior to acquisition and, more important, to preclude the possibility that the automatic orientation sequence may have terminated prematurely. The third in the series of preparatory commands was Undervoltage Protection On, but this was sent only if analysis by the Spacecraft Analysis and Command (SPAC) Group (located at the SFOF during launch) was confident that the spacecraft power level was normal and that the spacecraft was operating properly. Following the spacecraft's execution of Undervoltage Protection On, the Pioneer was ready for experiment turn on and the all-important Type-II orientation maneuvers.

Experiment Turn-On

The Pioneer scientific instruments were turned on by command during the first pass over Johannesburg (DSS–51). The planned sequences for the Block-I and Block-II spacecraft are indicated below. Usually, experiment turn-ons were separated by other spacecraft status commands, instrument calibration commands, and, in the case of Pioneer 6, the partial Type-II orientation commands. The actual events are described later for each mission in chronological order.

Experiment Turn-On Sequences

Block I	Block II
Ames plasma	Goddard cosmic dust
Goddard magnetometer	Ames plasma
MIT plasma	Ames magnetometer
Chicago cosmic ray	TRW electric field
GRCSW cosmic ray	GRCSW cosmic ray
Stanford radio propagation	Stanford radio propagation

The reasoning behind the above sequences was that those instruments measuring important near-Earth phenomena, particularly in the transition region as the spacecraft passed through the magnetosphere, should be turned on and calibrated first. Generally, about 20 min were scheduled between each experiment turn on. The Stanford Radio Propagation Experiment, which required the transmission of signals from Stanford University at Palo Alto, was usually not turned on until just before the first Goldstone (DSS–12) pass.

Orientation Maneuvers

The purpose of the Type-II maneuver was the rotation of the spacecraft spin axis about the Sun line until the spin axis was perpendicular to the plane of the ecliptic. As explained more fully in Vol. II, this maneuver was normally controlled from Goldstone where the Operations Orientation Director (OOD) maximized the telemetry signal received from the Pioneer's high-gain telemetry antenna. Generally, hundreds of Type-II orientation commands were relayed to the spacecraft, each giving rise to a pulse of gas from the orientation subsystem. Usually, there was some jockeying back and forth across the peak in the signal-strength-reception curve. On occasion, the normal Type-II orientation process was interrupted for another Type-I maneuver to remove any spin-axis misalignment inadvertently introduced by cross coupling during Type-II maneuvers. Often, orientation maneuvers were commanded even after the beginning of the cruise phase to "trim" spacecraft altitude and correct for drift, solar pressure effects, and other perturbations.

Preliminary trajectory analysis in the cases of Pioneers 6 and 9 indicated that the so-called "partial Type-II orientation" would be desirable early in the flight to preclude an unfavorable spacecraft orientation later in the flight. As discussed in Ch. 4, Vol. II, the low-gain omnidirectional antenna used for communication early in the flight had a very low gain within a 10° cone aft along the spin axis. During the partial Type-II orientation maneuver, the gas pulses torqued the spin axis sufficiently so that Goldstone antennas would not be looking up this cone at the spacecraft during the final Type-II orientation maneuver.

The partial Type-II orientation maneuvers were performed during the first passes of the spacecraft. For Pioneer 6, Johannesburg (DSS–51) was responsible for this special maneuver (table 4–1). But on the Pioneer-9 flight, it was decided to wait 4 hr until the spacecraft had been acquired by Goldstone (DSS–12), where the OOD and his team were already situated for the final Type-II orientation maneuvers, which customarily took place a pass or two later over Goldstone.

The final Type-II orientation maneuvers were always directed from Goldstone. Special equipment for this task as well as the OOD and his team were positioned here. The OOD began this maneuver when control was transferred to him from the Space Flight Operations Director (SFOD). His first commands were: (1) Format C, 512 bps, which provided maximum engineering telemetry to check spacecraft status and (2) Type-I orientation, to trim the orientation with respect to the Sun line. If all seemed to be going well, the command was given to reduce the engineering telemetry rate to 16 bps so that no data would be lost if the high-gain antenna was switched on and it happened to have Goldstone in one of the minima of its lobed gain pattern. (See antenna patterns in Ch. 4, Vol. II.) Next, the OOD commanded the spacecraft

TABLE 4–1.—*Deep Space Stations Participating in Orientation Maneuvers*

Orientation maneuvers	Pioneer 6	Pioneer 7	Pioneer 8	Pioneer 9
Initial acquisition	DSS–51[a]	DSS–51	DSS–41	DSS–51
Initial Type-I orientation [b]	DSS–51	DSS–51	DSS–41	DSS–51
Partial Type-II orientation	DSS–51	—	—	DSS–12
Final Type-II orientation	DSS–12	DSS–11	DSS–12	DSS–12

[a] DSS–11, Goldstone; DSS–2, Goldstone; DSS–41, Woomera; DSS–51, Johannesburg.
[b] The first Type-I orientation is automatic and occurs before and during DSS acquisition.

to switch from the low-gain omnidirectional antenna to the high-gain directional antenna. This antenna had to be used if telemetry was to be received from deep space. The spacecraft was now ready for the final Type-II orientation commands.

By direction of the OOD, blocks of Type-II orientation commands were transmitted to the spacecraft. Because each separate command released a single gas pulse which torqued the spin axis only about 0.3°, blocks of up to 30 commands had to be used to achieve noticeable changes in the signal strength detected by the Goldstone station. In practice, each block of commands was followed by a short period of analysis, during which primary interest was focused on the plot of received signal strength versus the number of Type-II commands transmitted. As further blocks of commands were sent, the plots would show definite minima and maxima characteristic of the Pioneer high-gain antenna. The major lobe was usually easy to recognize by its size. Nevertheless, further commands were issued beyond this peak until the signal strength had been roughly halved. This procedure insured that the main lobe had truly been found, and it permitted the OOD's engineers to compute the number of commands between the maximum and half-power point. It was then possible to return to the maximum by transmitting a fixed number of commands.

Interspersed with the above sequence of blocks of commands were occasional Type-I orientation commands which, as mentioned earlier, were necessary to reduce the effects of cross-coupling between the two degrees of freedom. These Type-I maneuvers were commanded only at maxima in the antenna pattern where the spacecraft telemetry could be safely switched to 512 bps. Only at 512 bps was status data transmitted faster than the automatically generated Type-I gas pulses.

A few reversals of the Type-II torquing sequence were also commanded to insure the OOD that the entire orientation subsystem was working satisfactorily. It would have been disastrous to reach the half-power point on the other side of the signal-strength maximum and find that backtracking to the maximum was impossible.

Usually, the Type-II maneuvers were interrupted several times for a half hour or more in order to assess the performance of the scientific instruments. Format B and the highest bit rate commensurate with the spacecraft signal strength were commanded during these periods.

Upon the successful completion of the Type-II orientation maneuver, the Type-I orientation command was given once more. Then, the spacecraft was placed in its cruise mode, with a telemetry bit rate of 512 bps, Format B. Spacecraft status was checked a final time, and, to terminate the maneuvers, the OOD sent the Orientation Power Off command. The

OOD then transferred mission control to SFOD at the SFOF, and the cruise phase began.

The Pioneer orientation maneuvers were unique. Despite considerable initial skepticism about the feasibility of the whole concept, the maneuvers proved relatively easy to carry out and control in practice. Other small and moderately sized spacecraft, American and foreign, have adopted similar altitude-control strategies.

Predicts

Upon acquisition, the SPAC group at the SFOF in Pasadena began to generate orbital data. The first orbit based on data from the first acquisition station was usually available about 1.5 hr after launch. Orbital data were teletyped to Ames Research Center where station look angles were computed for DSS–41, DSS–51, and DSS–12. Earth-spacecraft-Sun angle computations were also computed and teletyped to the SFOD and the OOD at Goldstone.

PIONEER OPERATIONS—ACQUISITION TO CRUISE PHASE

The Pioneer flights generally adhered to the scenario just described. Each, however, was different in several ways. The best technique to describe these specific differences and, in addition, convey the flow of events, is by chronologically recording the highlights of each flight from DSS acquisition to the beginning of the cruise phase. Tables 4–2 to 4–5 assemble these important events. For the purpose of illustration, Pioneer 6 events are presented in more detail than the other three flights.

TABLE 4–2.—*Highlights of the Pioneer-6 Flight: Acquisition to Cruise*

Date, GMT	Event and remarks
Dec. 16, 1965	
0759	Initial downlink acquisition by DSS–51.
0800	First telemetry signals indicated that the Type-I orientation was in progress and spacecraft normal.
0804	Type-I orientation completed. An estimated 386 ± 6 gas pulses were required, corresponding to a rotation of $64.1° \pm 1.0°$.
0813	Type-I orientation command given. No gas pulses indicated by telemetry inferring that the automatic Type-I maneuver had been successful.
0824	Undervoltage Protection On command transmitted.
0825	Initial downlink acquisition at DSS–42.
0835	The signal received from the spacecraft began to fade rapidly and "lock" was lost by the computer. Telemetry indicated that a signal was present in both receivers. Although both spacecraft failure and poor spacecraft orientation

TABLE 4–2.—*Highlights of the Pioneer-6 Flight: Acquisition to Cruise*—Continued

Date, GMT	Event and remarks
	were suspected at first, the problem was finally attributed to a normal but unforeseen phenomenon of coherent operations.
0851	First of 33 counterclockwise Type-II orientation commands sent from DSS-51 as part of partial Type-II orientation maneuver. No change in signal strength was noted. The inference from the spacecraft antenna pattern was that the spin axis had changed either from 125° to 135° or from 60° to 70° with respect to the North Ecliptic Pole.
0914	Ames Plasma Experiment turned on.
0931	First of 32 additional Type-II orientation commands sent.
0957	MIT Plasma Experiment turned on. Received signal strength increased 2 dB, assuring the OOD that the first change had been from 125° to 135°. With the spacecraft axis now between 140° and 150°, the partial Type-II orientation maneuver was terminated.
1013	Format-A command executed.
1031	Chicago Cosmic-Ray Experiment turned on.
1050	GRCSW Cosmic-Ray Experiment turned on.
1110	Stanford Radio Propagation Experiment turned on.
1130	Type-I orientation command sent. Single gas pulse indicated.
1255	Spacecraft penetrated magnetosphere 12.8 radii from Earth.
1710	Earth-solar wind bow shock penetrated at 20.5 Earth radii.
1912	First Stanford radio propagation data on teletype.
2003	First acquisition by DSS-12.
2100	First command link transfer, DSS-51 to DSS-12.
Dec. 17, 1965	
2008	Second acquisition by DSS-12.
2120	Control transferred from SFOF to DSS-12 for completion of Type-II orientation maneuver. Also, first data sent to experimenters from Ames Tape Processing Station.
2128	Type-I orientation command sent. Seven gas pulses counted.
2130	Telemetry changed to 16-bps mode.
2137	Spacecraft commanded to Format C, 16 bps.
2158	Spacecraft transmitter switched to high-gain antenna.
2210	Block of 33 counterclockwise Type-II orientation commands sent 1 min apart.
2250	Telemetry bit rate changed to 512 bps.
2254	Type-I orientation command given to correct for cross coupling. Four pulses noted from spacecraft telemetry.
2321	Block of 67 counterclockwise Type-II orientation commands dispatched 1 min apart.
Dec. 18, 1965	
0031	Ten more Type-II commands sent.
0048	Four gas pulses noted when Type-I orientation sequence was repeated.
0114	Block of 15 counterclockwise Type-II orientation commands sent.
0137	Only one gas pulse counted when Type-I orientation sequence was repeated.

TABLE 4–2.—*Highlights of the Pioneer-6 Flight: Acquisition to Cruise*—Concluded

Date, GMT	Event and remarks
0144	Block of 3 clockwise Type-II orientation commands sent to assure OOD that backtracking to maximum antenna lobe was possible.
0148	Open-ended block of counterclockwise Type-II orientation commands started. A total of 27 were sent.
0220	Three gas pulses noted as Type-I orientation sequence was repeated.
0223	Block of 10 clockwise Type-II orientation commands sent.
0345	Second block of 10 clockwise Type-II orientation commands sent.
0409	One gas pulse noted when Type-I orientation sequence was repeated.
0411	Orientation Electronics Off command executed.
0413	Format-A command executed.
0424	Mission Control returned to SFOF. Orientation maneuvers over; cruise phase began.

TABLE 4–3.—*Highlights of the Pioneer-7 Flight: Acquisition to Cruise*

Date, GMT	Event and remarks
Aug. 17, 1966	
1548	Initial acquisition by DSS-51. First telemetry indicated that Type-I orientation maneuver was in progress and that the spacecraft was functioning normally.
1557	Coherent, two-way lock established by DSS-51.
1600	Bit rate commanded from 64 to 512 bps.
1611	First acquisition by DSS-42.
1621	Signal received from spacecraft dropped rapidly.
1625	Due to ground station problems at DSS-51, the command link was transferred to DSS-42. Apparently, DSS-51 was tracking Pioneer 7 on a sidelobe of the ground antenna.
1702	Spacecraft entered Earth's shadow as indicated by bus-voltage telemetry.
1739	Spacecraft emerged from Earth's shadow.
1746	Battery on spacecraft turned off for 11 min due to high charging current overheating partially discharged battery.
1810	Undervoltage Protection On command transmitted.
1815	Experiment turn ons began. Completed at 2015.
Aug. 18, 1966	
0426	First acquisition by DSS-11. (DSS-12 was engaged in tracking other spacecraft.)
Aug. 19, 1966	
0429	Second acquisition by DSS-11. Type-II orientation maneuvers commanded from DSS-11 during second pass. Required 191 Type-II commands.

TABLE 4–4.—*Highlights of the Pioneer-8 Flight: Acquisition to Cruise*

Date, GMT	Event and remarks
Mar. 29, 1968	
1443	DSS–51 acquired first spacecraft telemetry. Telemetry indicated that all spacecraft subsystems were performing normally.
1455	Initial acquisition at DSS–41.
1457	DSS–41 acquired one-way lock.
1535	Spacecraft commanded from 64 to 512 bps.
1538	Telemetry indicated periodic fluctuation in primary bus voltage.
1550	Goddard Cosmic Dust Experiment turned on. By 2030, all experiments were on.
	Orientation power turned on during third pass over DSS–12.
April 1, 1968	
1900	Orientation power on.
	Type-I orientation maneuver performed automatically. Twenty-eight gas pulses noted.
1945	Type-I orientation maneuver commanded. Two valve pulses noted.
2000	Another Type-I orientation maneuver commanded. Only one valve pulse this time.
2030	Type-II orientation counterclockwise command sent. No valve pulse recorded.
2045	Type-II orientation clockwise command sent. One valve pulse recorded, as expected.
2100	Again a counterclockwise valve pulse was attempted, and again the telemetry showed that none actually occurred.
2115	Type-II orientation counterclockwise command sent through other spacecraft decoder. Still the valve did not pulse. The implication was that the Sun-sensor thresholds had degraded just as they had on Pioneers 6 and 7 despite the thicker cover glasses tried on Pioneer 8. (See Ch. 4, Vol. II.)
2130	Type-I orientation command dispatched. Telemetry indicated a single valve pulse.
2145	Command sent to turn off orientation power and enter cruise mode.

TABLE 4–5.—*Highlights of the Pioneer-9 Flight: Acquisition to Cruise*

Date, GMT	Event and remarks
Nov. 8, 1968	
1012	Momentary signal picked up at DSS–51.
1024	Initial acquisition by DSS–51. Acquisition delayed by a malfunctioning driver. Predicts were also late. First telemetry (one-way mode) indicated spacecraft systems were all performing normally.
1030	Two-way coherent mode established between DSS–51 and spacecraft.
1045	Spacecraft commanded to switch from 64 to 512 bps.

TABLE 4-5.—*Highlights of the Pioneer-9 Flight: Acquisition to Cruise*—Concluded

Date, GMT	Event and remarks
1046	Type-I orientation maneuver commanded.
1047	Three-way lock with DSS-51 and DSS-42 established.
1100	Undervoltage Protection On command sent.
1130	First experiment turned on. On the first pass over DSS-12, a partial Type-II orientation maneuver was performed. A total of 58 counterclockwise pulses were transmitted, rotating the spacecraft spin axis an estimated 15°.
Nov. 9, 1968	
2215	First of the final Type-II orientation commands were transmitted from DSS-12. A total of 250 counterclockwise pulses were commanded before the signal maximum was reached. Three test clockwise pulses were initiated to insure that the return to signal maximum could be made. A total of 26 clockwise pulses brought the spin axis back to maximum after the 6.7-dB point beyond the maximum was reached.
Nov. 10, 1968	
0600	Type-II orientation complete. Pioneer 9 enters cruise mode.

CHAPTER 5

Spacecraft Performance During the Cruise Phase

THE PIONEER SPACECRAFT WERE DESIGNED for a minimum life of 6 months each, and each greatly exceeded this goal. In fact, each spacecraft functioned well for several years, confirming in their longevities the design decisions made by Ames and TRW Systems in the early 1960's. This chapter is concerned with spacecraft performance in orbit around the Sun: How did each subsystem perform in practice? What components and design features finally encountered trouble? This introspection is well worthwhile because the basic Pioneer design philosophy has proved so successful that much of it is being applied to other spacecraft destined for the outer planets, such as the Jupiter fly-by probes, Pioneers F/G, which require years rather than months of successful operation.

As a chronological framework for the following discussion of spacecraft performance, table 5–1 provides a list of major engineering and scientific events during the cruise phase. An engineering event, of course, is one involving subsystem performance; say, the failure of a key component. Because improvements in ground support equipment have been so critical to extending Pioneer operation to greater and greater distances, some of these changes are also presented in the chronology. A scientific event might be a solar flare or the passing of the spacecraft behind the Sun. Thus, table 5–1 also identifies astronomical events of significance to the next chapter which deals with scientific results.

PIONEER-6 PERFORMANCE

The nominal Pioneer-6 mission extended from the launch date (December 16, 1965) to June 13, 1966—a total of 180 days. However, because spacecraft performance at the end of 180 days continued to be good and the 210-ft dish at DSS-14 became available for long distance tracking, the mission was extended. (All subsequent Pioneer missions were also extended, as spacecraft lifetimes greatly exceeded the 180-day design level.)

Although each Pioneer surpassed the goals set for it, each of the spacecraft provided its share of minor problems. On Pioneer 6, for example,

TABLE 5–1.—*Major Events During the Cruise Phases of Pioneers 6, 7, 8, and 9*

Date, GMT	Event
Pioneer 6	
Dec. 23, 1965	First Goddard magnetometer flip command executed.
Dec. 24, 1965	First Duty Cycle Store mode commanded due to first lack of DSS coverage.
Jan. 3, 1966	First acquisition by a DSS station without GOE equipment; DSS–61 (Robledo. Spain).
Jan. 13, 1966	Spacecraft control shared between Ames and SFOF for the first time.
Feb. 23, 1966	Transfer of mission control to Ames completed.
Mar. 2, 1966	Inferior conjunction or syzygy, with spacecraft 1.84° below Sun as seen from Earth.
Mar. 17, 1966	Bit error rate at DSS–12 reaches 10^{-3}; spacecraft bit rate reduced from 256 to 64 bps.
Apr. 13, 1966	Bit rate reduced from 64 to 16 bps. Pioneer 6 34.2 million km from Earth.
Apr. 29, 1966	Spacecraft receiver 2 switched to high-gain antenna.
May 8, 1966	Bit rate reduced from 16 to 8 bps. Pioneer 6 55.3 million km from Earth.
May 20, 1966	Perihelion; spacecraft 64 701 502 km from Earth, 121 821 430 km (0.814 AU) from Sun.
June 4, 1966	Type-I orientation maneuver; number of gas impulses indeterminate (4 to 10). Type-II orientation maneuver to confirm spacecraft altitude.
June 9, 1966	Type-II orientation maneuver for the purpose of attaining a more favorable spacecraft altitude for the extended mission. Battery switched off.
July 11, 1966	Stanford radio propagation experiment turned off because spacecraft was too far away.
Dec. 16, 1966	Magnetometer flipped by command.
Nov. 23, 1968	Superior conjunction. Excellent scientific data acquired as spacecraft passes behind the Sun.
Nov. 28, 1969	First simultaneous tracking of Pioneers 6 and 7 in "radial-spiral" experiment.
July 6, 1970	Magnetometer data all zeros.
Oct. 26, 1970	First simultaneous tracking of Pioneers 6 and 8 in "radial-spiral" experiment.
Oct. 30, 1970	Bit rate of 64 bps now standard.
May 19, 1971	Pioneers 6 and 8 aligned with Earth.
Pioneer 7	
Aug. 25, 1966	Magnetometer flipped by command for first time.
Aug. 31, 1966	TWT 2 switched in to replace erratic TWT 1.
Sept. 20, 1966	Syzygy, Pioneer 7 enters Earth's magnetic tail.
Nov. 6, 1966	Went to 64 bps because of high error rate.
Jan. 19, 1967	Lunar occultation pass from DSS–12.
Mar. 16, 1967	Type-II orientation maneuvers.

TABLE 5–1.—*Major Events During the Cruise Phases of Pioneers 6, 7, 8, and 9*—Concluded

Date, GMT	Event
Mar. 21, 1967	Bit-error rate of 10^{-3} reached with 85-ft antenna. Goldstone tracking continued using narrowed bandwidth and lowered noise temperatures.
June 13, 1967	Battery commanded off.
Aug. 17, 1967	Type-II orientation maneuvers.
Jan. 23, 1968	DSS test with linear polarization cone.
Feb. 16, 1969	Lack of Sun pulses noted in spacecraft telemetry. These returned later.
May 7, 1969	Undervoltage problems. (See text.)
Nov. 10, 1969	Aphelion.
Nov. 28, 1969	First simultaneous tracking of Pioneers 6 and 7 in "radial-spiral" experiment. Battery turned off.

Pioneer 8

Date, GMT	Event
Jan. 17, 1968	Syzygy, Pioneer 8 in Earth's tail.
Jan. 25, 1968	Magnetometer flipped by command for first time.
Jan. 27, 1968	Spacecraft emerges from geomagnetic tail.
Feb. 9, 1968	Type-I and partial Type-II orientation maneuvers.
Mar. 30, 1968	Type-I and Type-II orientation maneuvers revealed that Sun-sensor D was inoperative. Battery turned off.
June 27, 1968	Another orientation maneuver attempt showed that Sun-sensors A, B, and C were also out of action.
Sep. 21, 1969	Battery commanded off.
Jan. 20, 1970	Began electromagnetic interference tests to determine whether Goddard cosmic-dust experiment was being affected by spacecraft.
Oct. 26, 1970	First simultaneous tracking of Pioneers 6 and 8 in "radial-spiral" experiment.
May 19, 1971	Pioneers 6 and 8 aligned with Earth.

Pioneer 9

Date, GMT	Event
Jan. 14, 1969	Solar-array temperature on this inbound flight had begun rising, causing the primary bus voltage to decrease. Battery was disconnected.
Jan. 30, 1969	Inferior conjunction or syzygy.
Feb. 5, 1969	Special test using Type-I and Type-II orientation commands showed that the spacecraft orientation subsystem was working well, inferring that the Sun-sensor ultraviolet filters had solved the Sun-sensor degradation problem.
Apr. 8, 1969	First perihelion at 0.754 AU, which was within the 0.8-AU design goal.
Jan. 20, 1970	Began electromagnetic interference tests to determine whether Goddard cosmic-dust experiment was being affected by spacecraft.
Sep. 29, 1970	Ames magnetometer turned off temporarily to eliminate 1.9-dB degradation of Stanford radio propagation experiment.
Dec. 18, 1970	Syzygy.

the gas leak in the orientation control subsystem caused some concern. The degradation of the Sun sensors plagued every Pioneer until Pioneer 9's ultraviolet filters finally solved the problem. Future designers of long-lived spacecraft should benefit from Pioneer experience; therefore, these summaries of engineering performance are organized on a subsystem basis.[6] Many of the observations concerning spacecraft design also apply to the rest of the spacecraft in the series.

Orientation Control Subsystem

The initial Pioneer-6 orientation maneuvers have already been described in Ch. 4. The orientation control subsystem operated flawlessly during these maneuvers and also during the attitude adjustments made in June 1966 to prepare the spacecraft for the extended mission. The fact that these maneuvers were executed successfully at the end of nominal life indicated that the gas leak that had developed during the launch phase did not compromise the mission at all. The shape of the gas pressure curve drawn from telemetered data (fig. 5-1) implies that the gas leakage rate was proportional to a pressure less than the bottle pressure itself. The inference is that gas escaped through the relief valve or a poorly sealed nozzle valve. At the conclusion of the June 9, 1966, maneuver, the gas pressure was approximately 100 psi, and leakage was apparently near zero. A reconstruction of the Pioneer-6 spin rate is provided in table 5-2. Apparently the gas leak was responsible for slight changes in the spin rate.

Thermal Control Subsystem

No problems arose with this subsystem; all temperatures were maintained well within the design limits. All temperature measurements gradually rose as the spacecraft approached perihelion 155 days after launch, falling slowly afterwards. Approximately one month after launch, predictions were made of spacecraft temperatures at 0.9 and 0.8 AU based upon known orbital conditions and the results of the thermal-vacuum tests. All except three parameters fell within 2.5 percent of the predictions (fig. 5-2). The nitrogen temperature was off 4 percent and the two Sun-sensor temperatures, 6 percent, at 0.814 AU. In addition, the upper part of the solar array ran at a higher-than-expected temperature. Minor discrepancies in spacecraft thermal-vacuum simulation and the protrusion of the Sun-sensor shields accounted for these larger-than-expected deviations.

---

[6] These performance summaries are based largely upon discussions in the TRW Systems final report: Pioneer Spacecraft Project, Final Project Report, Rept. 8830-28, Dec. 1969.

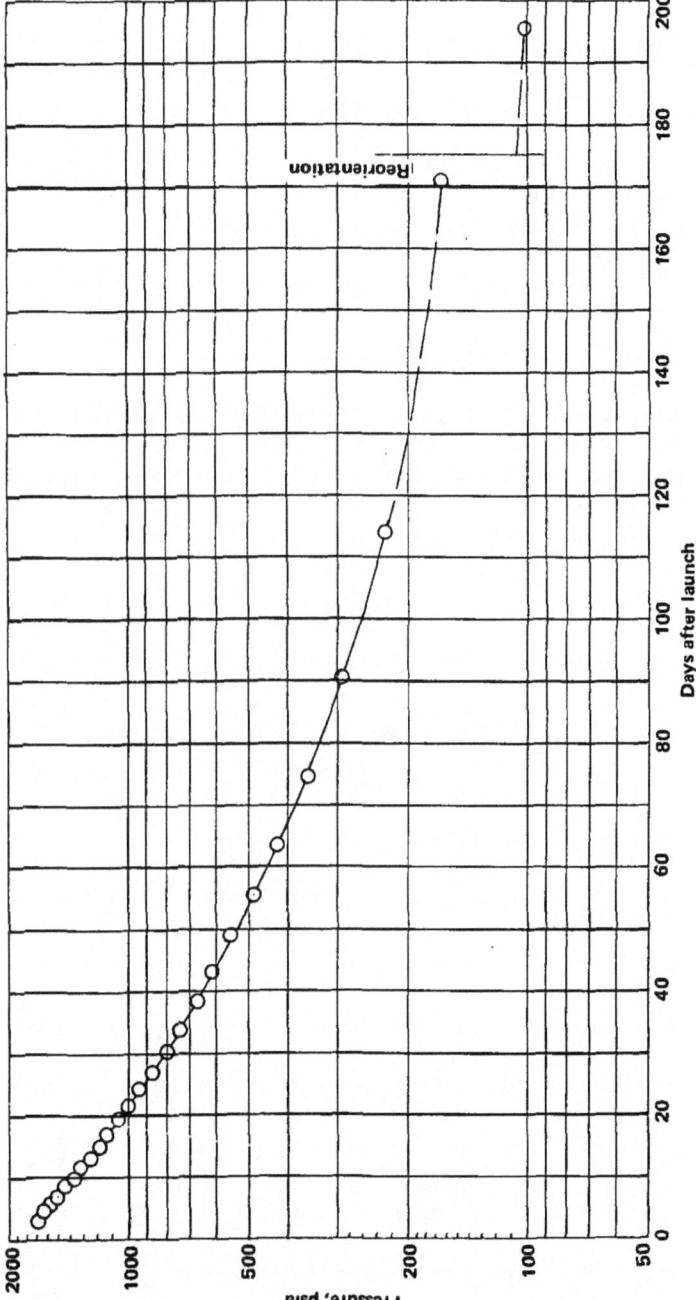

FIGURE 5-1.—Variation of Pioneer-6 nitrogen bottle temperature after initial orientation.

TABLE 5–2.—*Early Pioneer-6 Spin-Rate History*

Stage	Estimated spin rate, rpm
After boom deployment	58.88
After the automatic Type-I orientation	59.01
Before the partial Type-II orientation	58.99
After the partial Type-II orientation	59.11
Before the June 1966 maneuvers	57.77
After the June 1966 maneuvers	51.79

Electric Power Subsystem

Initial telemetry confirmed that the spacecraft power supply was operating normally (fig. 5–3). During the automatic Type-I orientation maneuver, the primary bus showed evidence of a ripple, but several groups of high-bit-rate, Format-C, telemetry data taken later in the revision did not indicate any ripple.

Right after boom deployment, the battery was recharged at an average 0.1 A. Fifteen min later, the charging current dropped to 0.024 A, and after 4 hr the battery was trickle charging or floating. On June 9, 1966, the battery was switched off the primary bus as a protective measure.

The spacecraft load history is summarized in table 5–3 and figure 5–4. Over the years the performance of the solar array has been degraded by solar particles, but this has not been considered a serious problem because of the favorable orbit.

TABLE 5–3.—*Pioneer-6 Electrical Load History*

Day of flight	Spacecraft condition	Volt	Primary bus Ampere	Watt [a]
1	Experiments off, orientation on	31.6	1.39	44.0
1	Experiments on, orientation on	31.1	1.63	50.7
2	Experiments on, orientation on	31.1	1.67	51.9
3	Experiments on, orientation on	31.1	1.67	51.9
3	Experiments on, orientation off	31.1	1.63	50.7
39	Experiments on, orientation off	31.1	1.63	50.7
53	Experiments on, orientation off	30.4	1.67	50.8
156	Experiments on, orientation off	28.2	1.79	50.5
171	Experiments on, orientation off	28.8	1.79	51.5
171	Experiments on, orientation on	28.8	1.83	52.6
208	Experiments on, orientation off	29.4	1.75	51.4

[a] Resolution was approximately ± 2 W.

SPACECRAFT PERFORMANCE 61

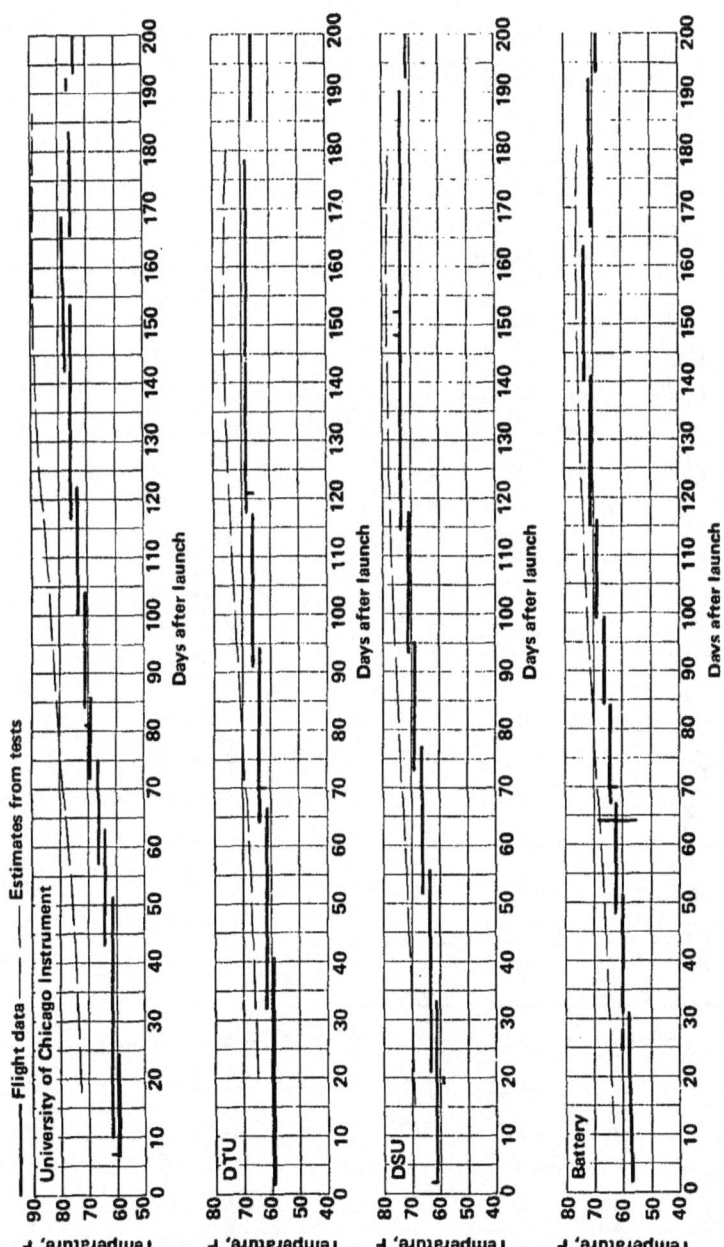

FIGURE 5-2.—Selected temperatures telemetered from Pioneer 6.

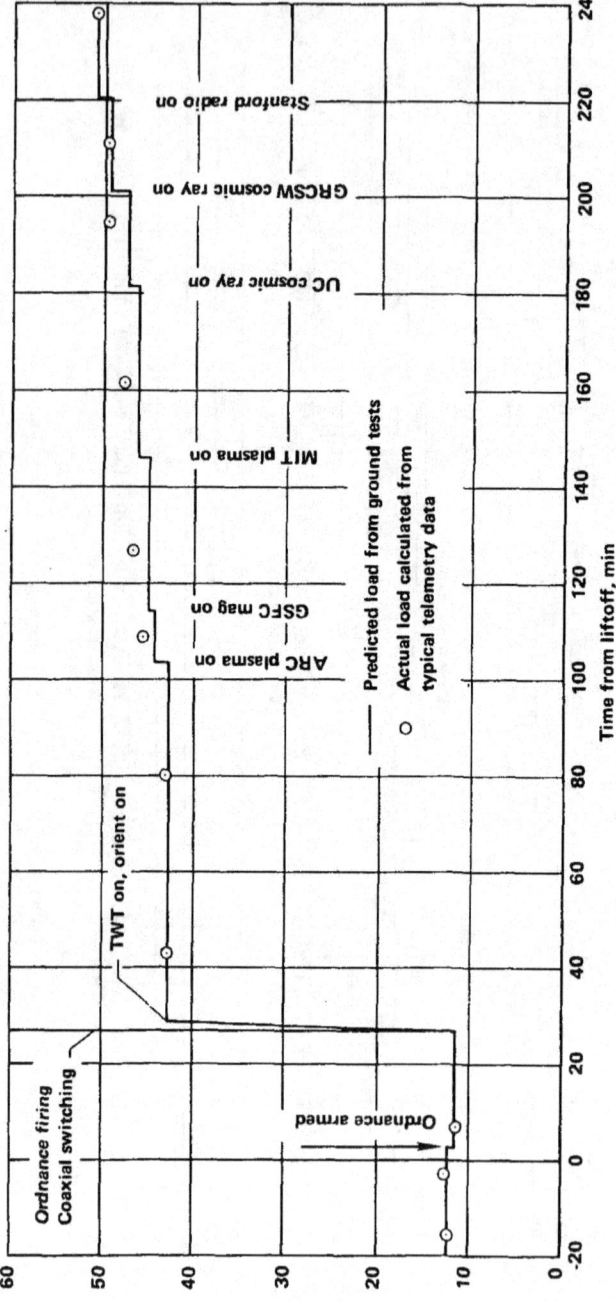

FIGURE 5-3.—Electrical loads on Pioneer 6 during the first 4 hr.

SPACECRAFT PERFORMANCE 63

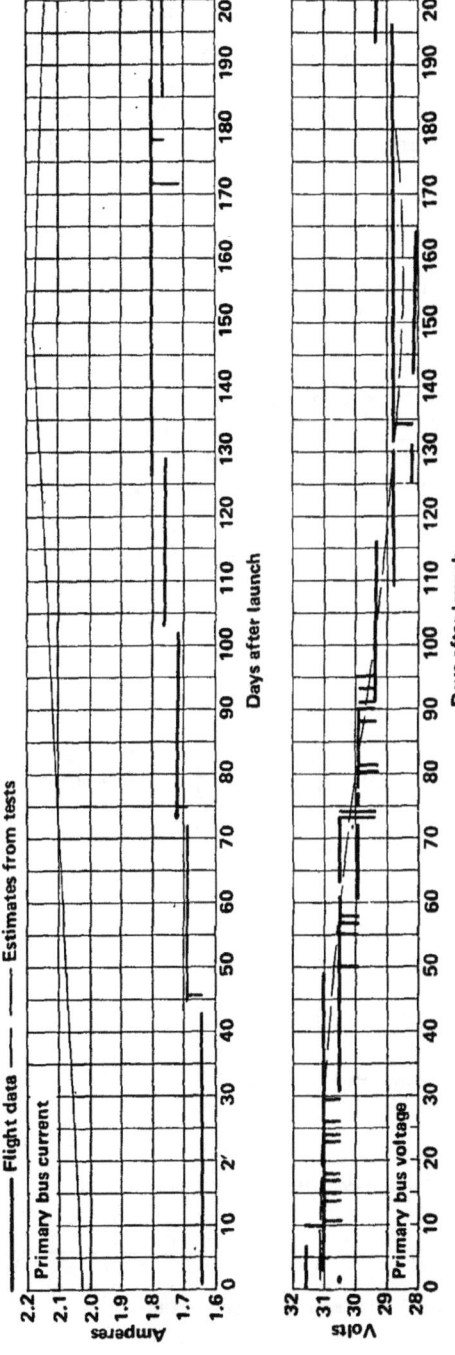

FIGURE 5-4.—Early history of the Pioneer primary bus voltage and current.

Communications Subsystem

The performance of this subsystem was generally better than predicted during the first 6 months of flight (fig. 5-5). The data in table 5-4 substantiate this observation for the March 1966 period. Figure 5-1 shows the actual bit-error rate for the same period. During these 6

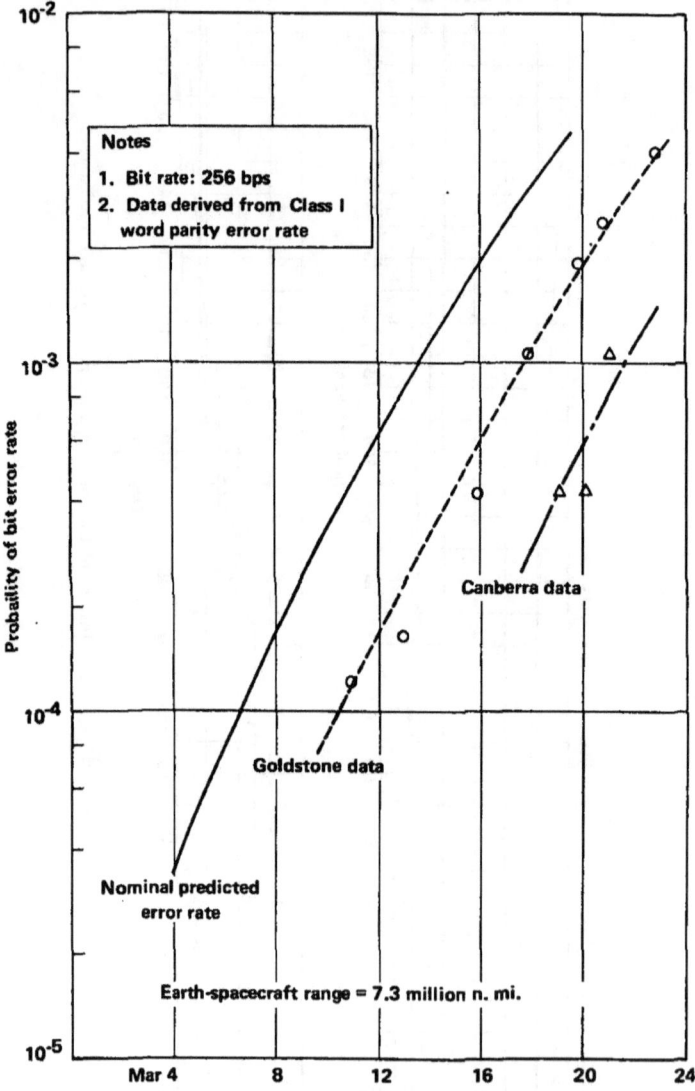

FIGURE 5-5.—Pioneer-6 telemetry bit-error rate during part of Mar. 1966.

months about 3500 commands from the Earth were executed satisfactorily by the spacecraft.

Housekeeping telemetry indicated a slight upward drift of the helix current of TWT 1 during the first 6 months. From 7 mA after initial stabilization, the current level crept up to between 7.50 and 7.75 mA during the first 175 days. In early 1971, the level had reached an 8.0-mA average. The TWT threshold lies within this range, but no further drift has been noted, and no degradation of TWT performance was apparent. The operating life of this TWT exceeded the life tests made by the manufacturer.

Inferior conjunction or syzygy occurred on March 2, 1966, at 0530 for Pioneer 6. As the spacecraft moved nearer the Sun, the bit-error rate rose as indicated in figure 5–6. The radio noise contributed by the Sun had, of course, adversely affected the signal-to-noise ratio and thus the bit-error rate. Signal deterioration was most severe within 2.5° of the Sun. By mid-1969, the high-gain antenna characteristics had degraded to a level equal to those of the low-gain antenna. A malfunction in receiver 2 prevented the use of channel 7.

Structure Subsystem

The spacecraft structure subsystem functioned perfectly during separation and boom deployment. The stability of the signal received by the DSS stations showed further that the spacecraft was aligned and balanced with high precision.

Data Handling Subsystem

During the first 6 months of operation, all of the various modes, formats, and bit rates were commanded for one reason or another (fig. 5–7). This subsystem responded properly to all commands. During the first 200 days of operation, the spacecraft procured nearly 3 billion bits of information for transmission to Earth (table 5–5). These bits fell in the following categories:

Category	Bits, millions
Science data	2060
Engineering data	160
Parity check bits	370
Data identification bits	260
Other	20
Total	2870

About 96 percent of the scientific data were transmitted in Format A, less than 4 percent in Format B, and less than 0.03 percent in Format D.

TABLE 5-4.—*Pioneer-6 Communication Subsystem Performance*

Date (1966)	Time after launch (days)	Representative received signal strength at Deep Space Stations —dBm						Predicted signal strength, —dBm
		DSS-11	DSS-12	DSS-41	DSS-42	DSS-51	DSS-61	
March								
8	82							148.7
9	83	149.3						148.9
10	84		148.0		146.0			149.1
11	85		148.5	149.0		148.6		149.3
12	86		148.0	149.3		149.2		149.5
13	87	149.2		149.5		150.0		149.7
14	88	150.2			148.0	149.2		150.0
15	89		149.0			149.5		150.2
16	90					150.2		150.4
17	91		151.6	150.6				150.6
18	92					150.8		150.8
19	93		150.4		153.3	150.5		151.0
20	94		150.7		151.6	151.2		151.2
21	95				151.1			151.4
22	96		151.0	151.8		152.3		151.7
23	97			152.0				151.9
24	98		150.8		150.1	152.2	150.2	152.1
25	99		152.7	152.2		153.5	150.7	152.3
26	100		152.9	153.4			150.9	152.5
27	101		152.5			153.8		152.7

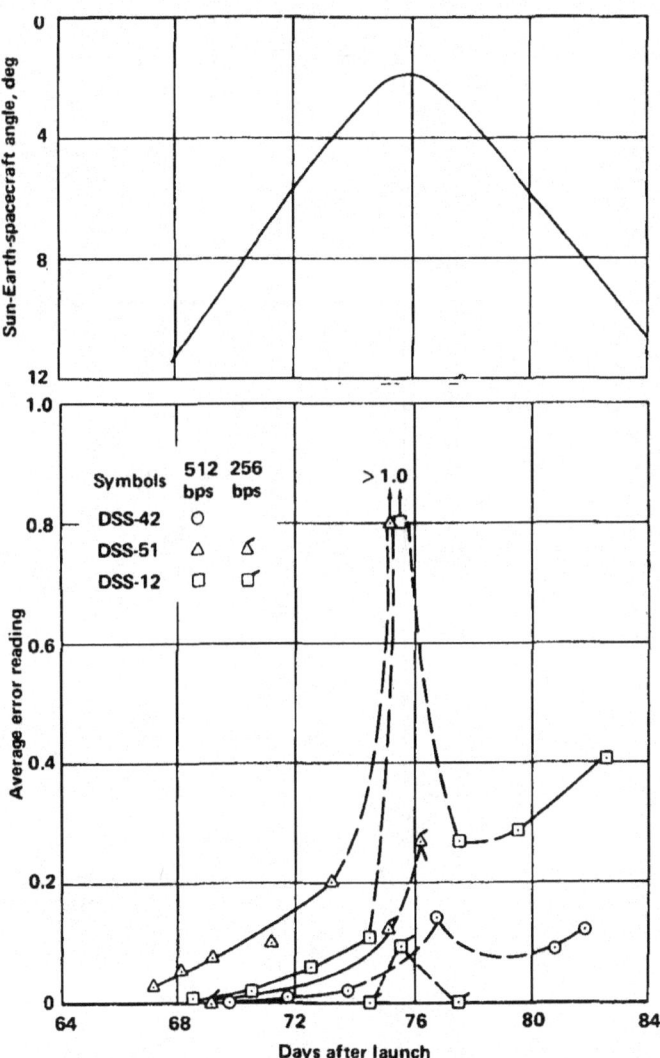

FIGURE 5-6.—The effect of proximity of Pioneer 6 to the Earth-Sun line of the telemetry data error rate.

The Real-Time Mode of data transmission was employed predominantly whenever DSS stations were available—in fact, almost all data received at Earth arrived via this mode. The Duty Cycle Store Mode provided about 18 percent of the data coverage, but because of intermittent sampling of stored data, this mode contributed less than 0.05 percent of the data received at Earth (fig. 5-8).

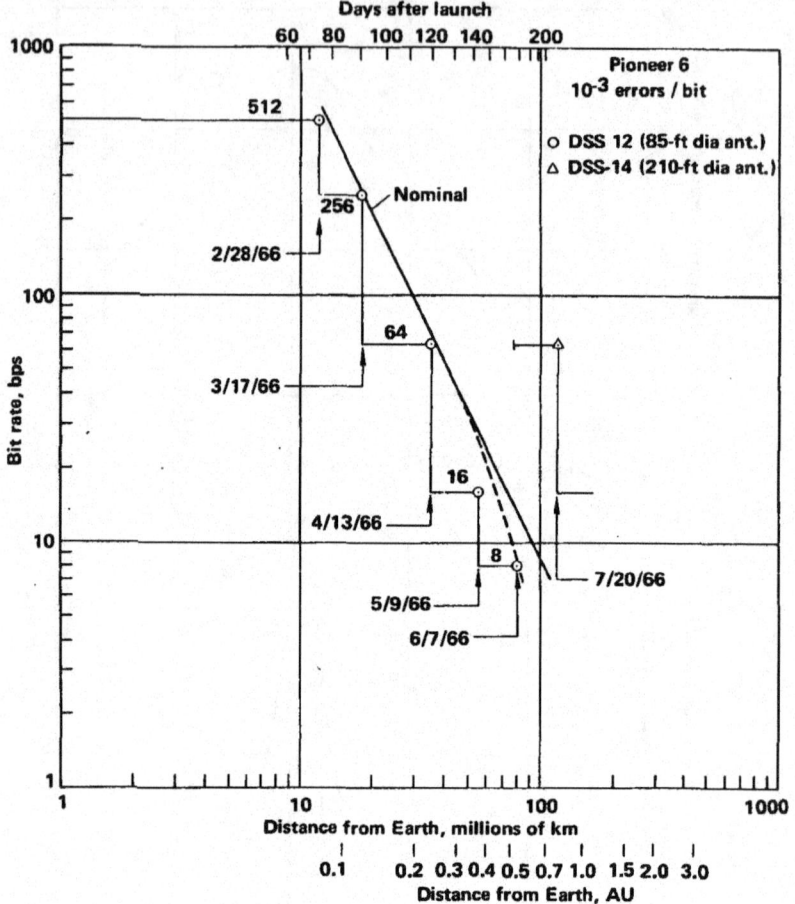

Figure 5-7.—Distance limitations for Pioneer-6 telemetry.

PIONEER-7 PERFORMANCE

As the spacecraft began the long cruise phase, all spacecraft subsystems appeared to be operating normally. On August 25, 1966, however, TWT 1 began to display anomalous performance in the noncoherent mode of operation although operation was normal in the coherent mode. For example, the helix current jumped to 10.2 mA compared to the nominal 6.1 mA, and the temperature rose to 180° F against the normal 101° F. On August 31, 1966, Ames personnel decided to switch in TWT 2. This TWT behaved normally in every respect. Except for this difficulty, overcome by redundancy in the design, spacecraft performance during the basic 180-day mission was excellent. Some subsystem details follow.

TABLE 5-5.—*Summary of Pioneer Data Acquisition Through 1969*[a]

Flight	Time in solar orbit, months	Billions of bits acquired	DSN telemetry support, hrs	Books of printed data[b]
Pioneer 6	48	3.030	6330	154
Pioneer 7	40	2.260	6271	118
Pioneer 8	25	6.032	4455	306
Pioneer 9	14	6.518	2275	332
Total	127	17.840 [c]	29173	910

[a] Adapted from Table 2, JPL Space Programs Summary 37-61, Vol. II, p. 13.
[b] Data printed in alpha-numeric form in 1000-page books.
[c] Of this total, 72 percent was scientific information, 6 percent engineering information, and 22 percent parity and data identification.

Orientation Subsystem

The orientation subsystem was turned off at 1102 GMT, August 19, 1966, after completing the usual series of Type-I and Type-II orientation maneuvers. Telemetry indicated a gas leakage rate of about 9 cc/hr, which was well within the specified maximum of 15 cc/hr. Some of these telemetered data are listed below. By 1970, essentially all of the nitrogen had leaked away (table 5-6).

On February 16, 1969, telemetry from the spacecraft indicated that the spacecraft was no longer generating Sun pulses. The precise time of failure is unknown, but it was between February 9 and 16. The opinion was that Sun-sensor E had degraded during the 914 days of flight to the point where the Sun no longer activated it. This failure was, of course, part of the ultraviolet degradation problem encountered with all Sun sensors except those of Pioneer 9 with ultraviolet filters. The lack of a reference Sun pulse has negated the magnetometer experiment and precluded anisotropy measurements with the GRCSW experiment.

Communication Subsystem

The performance of this subsystem has been excellent except for the difficulty of operating TWT 1 in the coherent mode, which was mentioned above. All five of the coaxial switches were activated once—successfully. Throughout the mission the redundant receivers were addressed occasionally; these, too, worked flawlessly. On January 4, 1967, receiver 2 was connected to the high-gain antenna. After January 10, all commands were transmitted to the spacecraft through this receiver.

It is interesting to compare the performance of the Pioneer-7 telemetry link with that of Pioneer 6. This is done in tables 5-7 and 5-8, by indicating the distance at which telemetry bit rate had to be reduced to keep the bit-error rate below 10^{-3}.

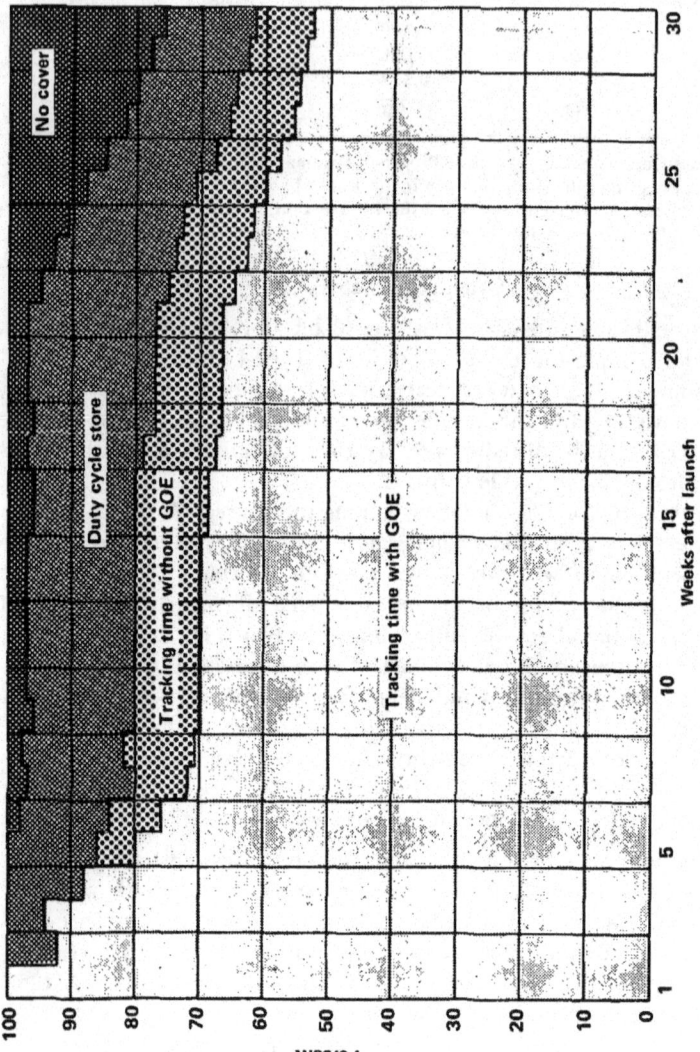

FIGURE 5-8.—Cumulative telemetry data coverage during the early part of the Pioneer-6 flight.

TABLE 5-6.—*Sun-Sensor E Degradation*

Date	Gas pressure, psia	Bottle temperature, °F
Aug. 19, 1966	2112	26.3
Sept. 2, 1966	2112 to 2048	23.9
Oct. 5, 1966	2048 to 1983	23.9
Nov. 10, 1966	1983 to 1918	23.9
Dec. 15, 1966	1918	21.5
Jan. 15, 1967	1853	21.5
Feb. 15, 1967	1789	19.1

TABLE 5-7.—*Comparative Telemetry Link Performance, Pioneers 6 and 7*

Threshold, bps	Pioneer-6 range, 10^6 km	Pioneer-7 range, 10^6 km	Predicted range, 10^6 km
512	12.4	12.2	12
256	18.3	15.0	17
64	34.3	32.3	33
16	57.0	59.1	60
8	79.5	76.8	80

TABLE 5-8.—*Received Carrier Strengths at DSS Receivers*

Range, 10^6 km	Pioneer-6 signal strength, dBm	Pioneer-7 signal strength, dBm	Predicted signal strength, dBm
1	−127.5	−127	−125.5
5	−139	−139	−139.4
10	−145	−145.5	−145.5
20	−151.5	−152	−151.5
40	−158	−158.2	−157.5
60	−162	−162	−161
80	−164	−	−163.5

Electric Power Subsystem

During all phases of flight the spacecraft power supply operated as predicted. During the eclipse of the Sun by the Earth, which began approximately 100 min after liftoff, the battery supplied 1.76 A. By the end of the eclipse, the solar-array temperature had dropped to −130° F and the bus voltage to 24.3 V. Entering sunlight again, the solar array picked up the load and began recharging the battery. The cold solar array generated a bus voltage of about 35 V and a current that saturated the current sensor at 0.562 A. To avoid overcharging the battery while the solar array warmed up, the battery was commanded off for 11 min. When the battery was reconnected, the voltage had dropped to 33.6 V and the recharging current to 0.130 A. Equilibrium conditions were gradually attained, and all parameters were normal.

On May 7, 1969, near the 1.125 AU aphelion, tests indicated that the undervoltage relay was being tripped when the MIT plasma experiment was switched to its high power mode. The implication was that the 9-W extra power requirement exceeded the capabilities of the degraded solar array. To provide the required power, the Goddard magnetometer and Stanford radio propagation experiments were turned off. This was acceptable because the magnetometer data were useless without the pulses from Sun-sensor E, which was out of action, and the spacecraft was beyond the range of the Stanford experiment. The MIT instrument was left in a low-power mode which prevented it from operating in its four highest energy steps.

Thermal Control Subsystem

Performance here has also been excellent. Temperatures of the Sun sensors did not decay as rapidly as prior analysis had indicated. This was attributed to the conservative approach used in the analysis. In this instance a conduction path to the spacecraft was not included in the analysis because it was to difficult to take into account.

During the ascent phase, the aft, uninsulated end of Pioneer 7 was illumined by the Sun. Platform temperatures during this phase were a few degrees higher than those of Pioneer 6 which was illumined on its insulated end. After the orientation maneuvers, Pioneer 7 continued to operate a little warmer as described below.

The equilibrium temperatures at various points on the spacecraft are listed in table 5–9 for the Pioneer-6 and Pioneer-7 flights. It is interesting to note that the Pioneer-7 platform ran a few degrees warmer than Pioneer-6 and that the solar-array temperature was a little lower. The inference is that Pioneer 7 was better insulated than Pioneer 6.

Structure Subsystem

Performance was good throughout the mission. The absence of ripple on the primary bus and signal stability at the DSS receivers inferred precision balance and alignment.

TABLE 5-9.—*Equilibrium Temperatures at 1 AU, Pioneers 6 and 7*

Telemetry measurement	Pioneer 6, °F	Pioneer 7, °F
Sun-sensor A	58	55
Sun-sensor B	80	82
Receivers	57	62
TWT 1 (operating)	103	106
TWT 2 (in reserve)	70	79
TWT converter	76	81
Transmitter driver	52	52
Digital telemetry unit	62	64
Data storage unit	64	67
Equipment converter	62	67
Battery	60	60
Upper solar panel	41	27
Lower solar panel	41	27
Platform 1	83	86
Platform 2	57	60
Platform 3	52	52
Wobble-damper boom	54	63
High-gain antenna	64	64
Lower actuator housing	73	79
Nitrogen bottle	27	28

Data Handling Subsystem

All formats, modes, and bit rates were used successfully during the mission.

PIONEER-8 PERFORMANCE

The Earth-escape hyperbola for Pioneer 8 was less energetic than planned. Instead of occurring at roughly 500 Earth radii, syzygy took place at 463 Earth radii. The heliocentric orbit was less eccentric and more inclined than the planned orbit, but the differences were not significant. The spacecraft performed normally except for the deviations noted below.

Early in the mission, trouble was experienced with the Ames plasma probe, and it was subsequently turned off. However, the difficulty was ultimately traced to a corona discharge resulting from outgassing. Later, the Ames experiment was switched back on, and it operated without further trouble.

Receiver 2 drifted somewhat, with a drift of 10 kHz being measured 3 months after launch.

During an orientation maneuver in March 1968, Sun-sensor D was found to be inoperative. On another orientation attempt in June 1968,

Sun-sensors A, B, and C were also found to be out of commission. The heavier Sun-sensor covers installed on Pioneer 8 had obviously not solved the degradation problem.

PIONEER-9 PERFORMANCE

Pioneer 9, an inbound flight, was subjected to increasing solar radiation, higher solar-array temperatures, and, consequently, falling bus voltages. To prevent the discharge of the battery, it was switched out on January 14, 1969.

To check the effects of the ultraviolet filters newly installed on the Sun sensors, a special test was conducted on February 5, 1969, the 89th day of flight. Telemetry indicated that Type-I and Type-II commands were executed properly. The ultraviolet filters had apparently solved the Sun-sensor degradation problem.

The spacecraft reached perihelion at 0.754 AU on April 8, 1969. The spacecraft was designed to penetrate to only 0.8 AU. It reached 0.754 AU without overheating although the cosmic-ray experiment reached 90° F, its upper limit.

All spacecraft systems operated normally throughout the nominal 180-day mission. During the extended mission in May 1969, the communication range reached 130 million km (78 million miles) using only the 85-ft DSS antennas, and a bit-error rate of 10^{-3}. This extension of the communication range can be attributed to three factors:

(1) Use of linear polarizers at some DSS stations
(2) Improvement of noise temperatures at the DSS stations
(3) Use of the Convolutional Coder Unit on Pioneer 9

Later in 1969, the communication range was extended still further to 152 million km (95 million miles) by allowing the bit-error rate to increase to 10^{-2}. During part of 1970, an improved, low-temperature cone (an "ultracone") was installed at DSS–12. With this improvement, the communication range was extended to 260 million km (162 million miles).

Decoder 2 began operating improperly in 1969 and is no longer used for normal operations.

Convolutional Coder Unit (CCU) Performance

The CCU, which is described in Vol. II, was added to Pioneers D and E as an engineering experiment. It can be switched in or out of the telemetry stream. CCU performance has been good, contributing about 3 dB to the communication power budget. In effect, the CCU has nearly doubled Pioneer-9's communication range.

Between the November 6, 1968, launch and December 10, 1968, the spacecraft operated in the uncoded mode at 512 bps except for CCU functional checks. Since December 10, the CCU has been in almost constant use except when the spacecraft is being worked by a DSS without Pioneer GOE.

About January 7, 1969, Pioneer 9 was far enough away for the CCU to provide a "coding gain" for DSS stations configured for receiving circularly polarized waves.[7] Up to March 6, 1969, GOE-equipped DSS stations tracked Pioneer 9 approximately 1000 hr with the CCU in operation; 680 of these hours were in the coding-gain region. As a result of the CCU's coding gain, 4.43×10^8 additional bits were received during this period. The 3-dB gain at 512 bps was verified by direct comparison with uncoded data at 256 bps. The CCU experiment has been so successful on Pioneer 9 that convolutional coding is being applied to other spacecraft.

REFERENCE

1. LUMB, D. R.: Test and Preliminary Flight Results on the Sequential Decoding of Convolutional Encoded Data from Pioneer IX. IEEE paper, International Conference on Communications, 1969.

---

[7] The Pioneers transmit linearly polarized signals. A loss is incurred when a DSS antenna receiving circularly polarized signals is used.

CHAPTER 6

PIONEER SCIENTIFIC RESULTS

THE COMPLETE SCIENTIFIC LEGACY of the Pioneer Program will not be known for many years. Scientific papers based upon the data telemetered back from deep space are still being published in abundance. Indeed, all four successfully launched spacecraft are still active and continue to add to our scientific store. The Pioneer scientific record of today, though incomplete, is impressive; some 137 contributions are listed at the end of this volume. These papers and some of their implications are summarized in the following pages.

The first Pioneer was launched in December 1965 to add to the substantial fund of information that Earth satellites and planetary probes had already discovered during the first 8 yr of the Space Age. It is impractical to review here the state of our knowledge of interplanetary space as of 1965. However, Glasstone's *Sourcebook on the Space Sciences*[8] was published about the same time as the launch of Pioneer 6, and the reader is referred to it for background information.

To set the stage properly, a brief description of the cosmic setting of the Pioneer drama is in order. The Sun controls most of what happens in interplanetary space. In 1965 solar activity was low and supposedly going to get even lower. Of course, the Pioneers were originally planned to support the world's IQSY program. However, solar activity, as measured by the sunspot number, began to climb in 1966; by 1969 it had reached its peak. The Pioneers, therefore, have already monitored solar activity in deep space for over half a solar cycle. As solar activity built up in the late 1960's, solar flares appeared more frequently, engulfing with their plasma and cosmic radiation some of the Pioneers and, on occasion, the Earth too. These flares were of great interest to science, and the Pioneers, located strategically around the Sun, in effect made the whole solar system a laboratory for Earth-bound scientists. The more important of these flares were noted in the chronology of the last chapter. Thus, the physical backdrop for the Pioneer program was one of increasing solar activity—more plasma-producing flares, more solar cosmic rays, and, in general, more opportunities to unravel the effects of the Sun on the interplanetary medium and the Earth.

---
[8] D. Van Nostrand, Princeton, 1965.

THE GODDARD MAGNETIC FIELD EXPERIMENT
(PIONEERS 6, 7, AND 8)

The magnetic field in interplanetary space is intimately associated with the hot, ionized plasma that streams outward from the Sun. In fact, it is customary to speak of the solar magnetic lines of force as "imbedded" in the plasma. Thus, data from the Pioneer magnetometers must be studied in conjunction with measurements made by the spacecraft's plasma instruments. The GRCSW cosmic-ray anisotropy experiment was also related to the magnetometer in the sense that cosmic-ray isotropy could be affected by changes in the structure of the magnetic field. In addition, on the Block-II Pioneers, the TRW Systems electric field experiment registers many of the same magnetohydrodynamic phenomena that are signaled by the magnetometer and plasma probes. The magnetometer therefore views only one dimension of the interplanetary plasma which, as we shall see, turns out to be a most complex medium indeed.

By December 1965 when Pioneer 6 was launched, satellites, such as IMP 1 (Explorer 18) and space probes, had already confirmed the theoretical prediction of a basically spiral solar magnetic field imbedded or "frozen" in the streaming solar plasma. The Sun's rotation about its axis imposed the so-called "water sprinkler" pattern on the outwardly rushing plasma (fig. 6–1). At the distance of the Earth, the solar plasma had a velocity of about 400 km/sec, and the water-sprinkler effect bent the solar magnetic lines of force until they were inclined about 45° to the Sun-Earth line. The IMP 1 magnetometer showed further that the direction of the solar magnetic field would be first directed outward and then inward, each condition lasting several days as the Sun's rotation carried the newly dubbed "sectors" past the Earth (fig. 6–1). It was first thought that the sectors might be associated with large plasma-emitting areas of the Sun, but some recent theories suggest that small nozzle-like regions may be responsible. The sectors also evolve with time (ref. 1); there were four separate sectors during the latter part of the 1960's, but their pattern varied. The Pioneer spacecraft were ideal platforms from which to monitor these gross structures of interplanetary space and also investigate any microstructure superimposed upon the sectors.

First Results from Pioneer 6

At approximately 11 Earth radii, the magnetometer reported a marked increase in magnetic field fluctuations as shown in figure 6–2 (ref. 2).[9] With the penetration of the magnetopause at 12.8 Earth radii, the field

---

[9] The first Pioneer-6 results were communicated at a special Pioneer-6 symposium convened by the American Geophysical Union in 1966.

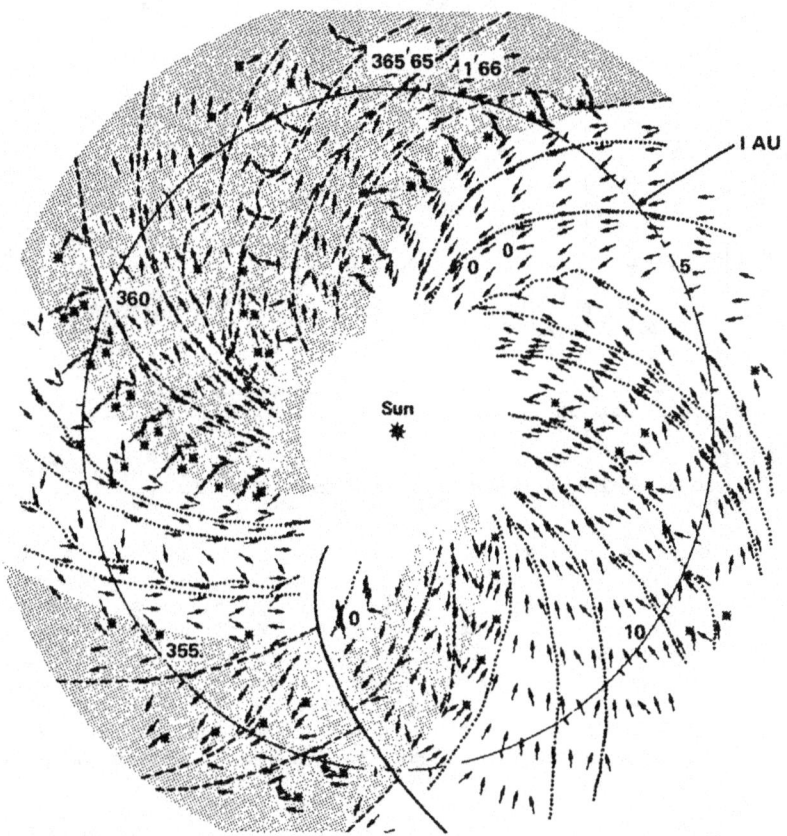

FIGURE 6–1.—Sector strucutre of the interplanetary magnetic field from Pioneer-6 data telemetered between Dec. 18, 1965, and Jan. 14, 1966. Each arrow represents an equivalent flux of 5 for 6 hr. Shaded regions are those where the field is directed away from the Sun; field was antisolar elsewhere. From: Schatten; et al. Solar Physics, vol. 5, 1968, p. 250.

dropped quickly to less than 20 γ–a well-known phenomenon by 1965. The collisionless bow shock was penetrated at about 20.5 Earth radii, and the quiet-time field then fell to about 4 γ.

Pioneer-6 data also confirmed that the interplanetary magnetic field often changes direction abruptly without changing magnitude. This phenomenon was interpreted at that time in terms of intertwined filamentary or tube-like structures in interplanetary space which mostly display the classical spiral structure but which sometimes create a twisted microstructure. The low-energy solar-proton anisotropies observed by Pioneer 6 tend to confirm this view (fig. 6–3).

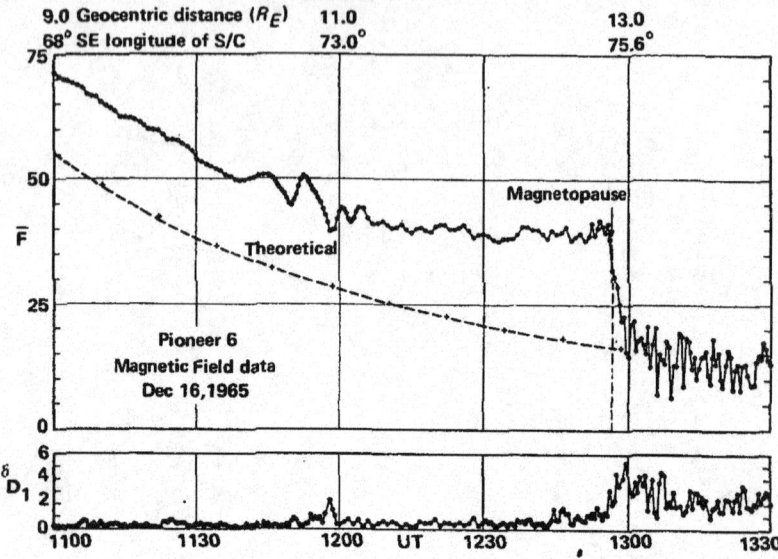

FIGURE 6-2.—Observations of geomagnetic field magnitude near the boundary of the regular field, the magnetopause, at a distance of 12.8 R_E near the sunset terminator. The observed magnitude is larger than the theoretically expected because of the compression of the Earth's magnetic field by the solar wind. Note the abrupt transition from strong and regular fields to weak and rapidly fluctuating fields. The lowermost "noise" curve measures the rms deviation (γ) over a 30-sec interval of one component of the magnetic field. Note that the increase in noise level as the magnetopause is approached and the significantly higher noise level when within the magnetosheath. From: ref. 2.

Pioneer-6 results were compared with magnetometer readings from IMP 3 (Explorer 28) (ref. 3). This was the first time that accurate measurements of the interplanetary magnetic field had been made from two widely separated spacecraft. By considering corotation delays (due to the Sun's rotation) excellent agreement was found between the two sets of data. Using these observations, Burlaga and Ness (ref. 4) identified a tangential discontinuity.

Generally, then, early Pioneer magnetometer data tended to confirm the Earth shock structure, the magnetopause, and the spiral sector structure of the interplanetary field inferred from previous spacecraft flights. The strong experimental support for a filamentary fine structure was perhaps the most interesting result from the first few months of flight.

Further Observations of the Geomagnetic Tail

Outward-bound Pioneers 7 and 8 carried Goddard magnetometers through the region where the geomagnetic tail was expected to exist.

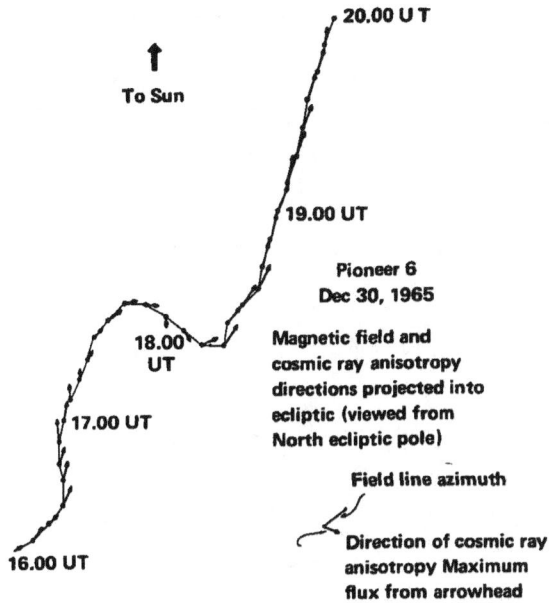

FIGURE 6-3.—Comparison of interplanetary magnetic field and solar cosmic-ray anisotropic directions projected onto the ecliptic plane. From: ref. 29.

This region was crossed by Pioneer 7 between September 23 and October 3, 1966, at distances ranging from 900 to 1050 Earth radii. Simultaneously, results from Explorer 33 demonstrated the existence of a tail out to 80 Earth radii. The Pioneer flights presented additional opportunities to explore this strange region "downwind" from the Earth.

A coherent, well-ordered geomagnetic tail with an imbedded neutral sheet was not observed by Pioneer 7 (ref. 5). However, the rapid field reversals recorded (fig. 6–4) are characteristic of the neutral sheet region observed closer to Earth. The conclusion of Ness and his colleagues at Goddard was that the geometry of the tail changes to a complex set of intermingled filamentary flux tubes at several hundred Earth radii.

In a later paper, Fairfield compared Pioneer-7 data with those obtained during the same period from Explorers 28 and 33 in the region closer to Earth (ref. 6). Comparisons revealed periods when Pioneer 7 was recording the steady, higher-magnitude solar or anti-solar fields characteristic of the tail but quite different from the fields connecting past the satellites. The presence of these isolated intervals was interpreted as due to the tail sweeping back and forth across the spacecraft in response to changing directions of plasma flow. Discontinuous features of the tail were found to connect past the three spacecraft at velocities comparable to the measured velocities of the solar wind.

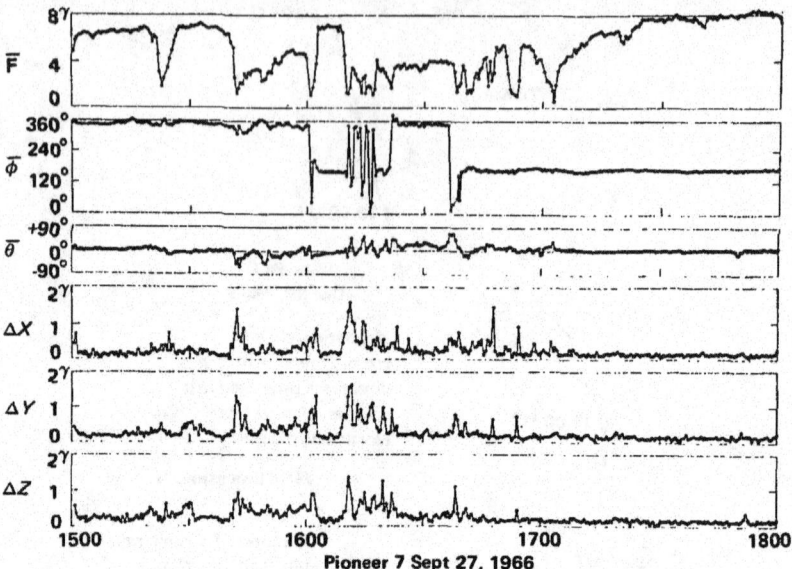

FIGURE 6-4.—Detailed 30-sec averaged magnetic field observations by Pioneer 7 on Sept. 27, 1966, when the field orientation and its rapid reversals are characteristic of the neutral sheet region of the geomagnetic tail. From: ref. 5. Throughout the 3-hr interval from 1500 to 1800 the field is observed to be directed either away from the Sun ($\phi = 180°$) or toward the Sun ($\phi = 360°$).

Pioneer 8 passed through the tail region at 470 to 580 Earth radii in January 1968. The results here were similar to those obtained from Pioneer 7 at 100 Earth radii (ref. 7). The geomagnetic tail may lose the clearcut structure plotted by Explorer 33 at 80 Earth radii before it reaches 500 Earth radii.

Mesoscale and Microscale Structures

Whereas the macroscale structures of the interplanetary field—those structures persisting 100 hr and more—generally fit theoretical expectations quite well, mesoscale structures (1 to 100 hr) and microscale structures (<1 hr) presented new experimental and theoretical problems and results. Some relevant observations and interpretations arising from Pioneer magnetometer data follow.

Directional discontinuities were correlated early with solar cosmic-ray anisotropies and explained in terms of spaghetti-like flux tubes or filaments. More thorough analysis of Pioneer data has replaced this type of model with a "discontinuous" model (ref. 8). The new model recognizes the fact that field discontinuities on the mesoscale and microscale—in both magnitude and direction—are more prevalent that previously suspected, and that their character does not always imply the existence of filaments.

In another paper, Burlaga (ref. 9) has reported a variety of other microscale structures:

(1) Transitional regions (called D-sheets) associated with plasma discontinuities

(2) D-sheets possibly related to the annihilation of magnetic lines of force

(3) Inhomogeneous, isothermal regions in which the square of the magnetic field intensity is proportional to the density and hydromagnetic tangential discontinuities in these regions

(4) Periodic variations in magnetic field intensity associated with discontinuities in the bulk speed

Burlaga suggested that small velocity discontinuities play a fundamental role in reducing stresses in the interplanetary medium, and that large velocity discontinuities may give rise to waves and turbulence.

Fluctuations and Power Spectra

The calculation of power spectra is useful in building theoretical models of the interplanetary milieu. For example, power spectra have been related to parameters describing the propagation of cosmic rays.

Power spectra based on early 1966 Pioneer-6 data exhibit an inverse dependence on the inverse square of the frequency as shown in figure 6–5 (ref. 10). The spectral shapes at that time were apparently dominated by microscale plasma-magnetic field discontinuities being connected outward by the solar wind. During more disturbed periods later in the flight, the discontinuities were not dominant, and the power spectra typically show an inverse 3/2 dependence on frequency.

Power spectra studies (ref. 11), using data obtained from Pioneers 7 and 8 as they passed through the magnetosheath, favor the view that fluctuations are created by the amplification of small connected irregularities occurring at the Earth's bow shock.

Observations of Specific Interplanetary Events

Pioneer magnetometer results have helped provide insight into what happens in interplanetary space when a major solar event, usually a large flare, occurs on the Sun. One such analysis concerned the July 7, 1966, event (ref. 12). The available Pioneer-6 information was used sparingly here in combination with Explorer-33 data. Pioneer data were also applied to the event of February 25, 1969 (ref. 13). The following

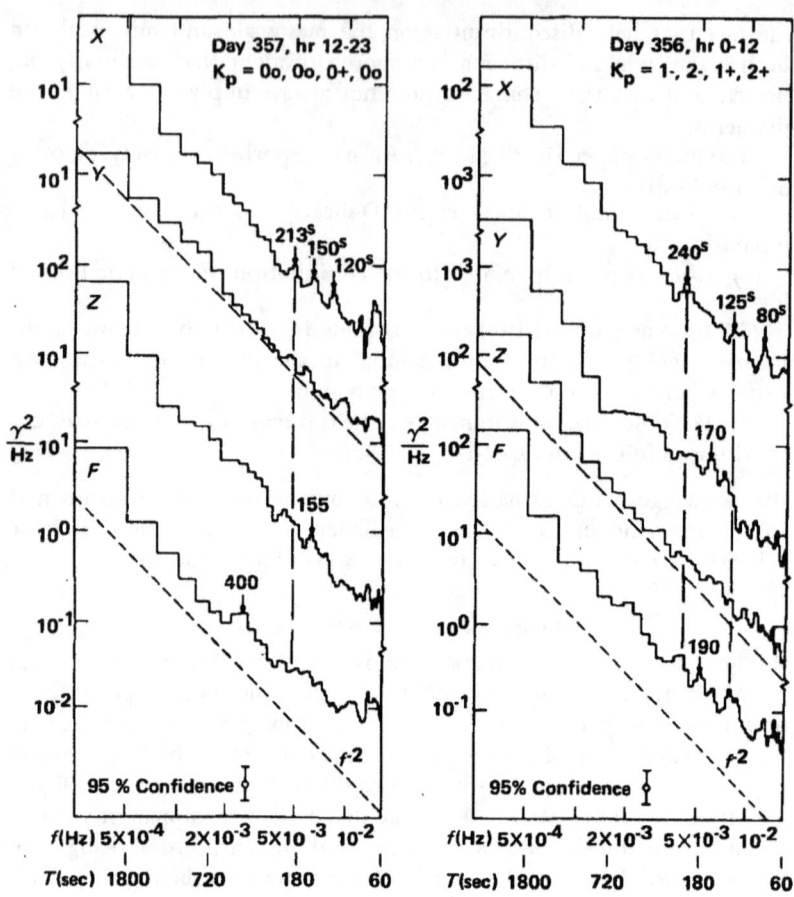

FIGURE 6-5.—Power spectra of the interplanetary magnetic field components and magnitude for two 12-hr periods in Dec. 1965. The dotted lines indicate inverse-square dependency. Data are from Pioneer 6, 10^6 km from Earth. From: ref. 10.

observations based on Pioneer-8 telemetry represent about what one would expect from a general model of a solar disturbance propagating through space.

(1) A rather steady field of 4 to 6 γ was observed during the early hours of February 25.

(2) The field increased rapidly to near 10 γ between 2000 and 2022, then it rose slowly to about 14 γ.

(3) Long-period variations were observed between 0200 and 0500, February 26.

(4) A very quiet field of about 6 γ occured between 2000 and 0500, February 27.

(5) The next group of telemetered data at 2149, February 27, again revealed a high field (over 10 γ). Large variations were noticed.

(6) In the last time interval telemetered, between 0200, February 28, and 0500, February 29, the field had dropped to normal values.

THE MIT PLASMA PROBE (PIONEERS 6 AND 7)

The preliminary MIT data presented at the American Geophysical Union Pioneer-6 Symposium indicated that, first, sharp changes in the plasma density preceded the dramatic changes in the magnetic field recorded by the Goddard magnetometer, and, second, the peaks in number density were followed by periods of increased bulk velocity (ref. 14). The early theoretical conclusions drawn from these coordinated measurements have already been covered in the preceding section.

The MIT group later published additional correlations between the plasma-probe and magnetometer data (ref. 15). In this study, the simultaneous changes in plasma and magnetic parameters were found to be consistent with what one would expect from tangential discontinuities. High-velocity shears were observed across these discontinuities, the largest being about 80 km/sec. The discontinuities observed by the MIT plasma probe were undoubtedly due to the same filament boundaries or discontinuities discussed in the papers published by the Goddard group.

The MIT plasma-probe and Goddard magnetometer data also showed that these discontinuities have preferred directions in space, with a tendency for the solar wind to be fast from the west and slow from the east. This east-west asymmetry in solar wind velocity is a natural result of the rotation of the Sun—the water-sprinkler effect again. Slow streams of plasma tend to spiral more tightly, and fast streams are straighter. The condition exists whereby slow and fast plasma streams tend to push against one another. Fast plasma streams push slow plasma away from the Sun and to the east; fast plasma, in turn, is pushed toward the Sun and to the west. The east-west asymmetry was shown strikingly when 3-hr averages of solar wind speed were correlated against flow angle in the plane of the ecliptic for a 27-day stretch of data (ref. 16). Figure 6–6 shows a conspicuous peak at zero lag, a positive correlation confirming the predicted in-phase relationship.

The correlation of plasma-probe and magnetometer data has continued to be a fruitful way to study the detailed structure of the solar plasma. For example, the general form of theoretical relation between the size of a sudden geomagnetic pulse and the associated change in solar wind stagnation pressure was confirmed in this way (ref. 17). Formisano was also shown that data from the two Pioneer instruments may be combined to study the mechanism that controls the high energy tail of the interplanetary electron distribution (ref. 18). It seems, for example, that electron pressures are usually two to five times higher than proton pressures.

Observation of Solar Flares

The MIT plasma probe, like the Goddard magnetometer, observed the passage of the shock front due to the solar flare of July 7, 1966 (ref. 19). The shock was first observed close to the Earth by the plasma probe and magnetometer on Explorer 33 some 45 hr after the visual observation of the solar flare. Ninety hours after the visual observation, Pioneer 6 recorded the passage of a shock. Because Pioneer 6 was closer to the Sun than Explorer 33, the anomalous time delays are difficult to explain. Even if the disturbance engulfed the Earth first, and then, through corotation, finally reached Pioneer 6, we have temporal inconsistencies. There may have been two separate interplanetary disturbances.

Observations in the Magnetosheath

Pioneer 6 carried the MIT plasma probe through the magnetosheath in the dusk meridian on December 16, 1965 (ref. 20). While the data confirmed some portions of the various theories developed to describe the magnetosheath, the proton distribution measured was bi-Maxwellian (fig. 6–7) rather than the classical single-peaked curve. Roughly 10 percent of the total number density was estimated to reside in the high-energy tail. The high-energy tail was observed throughout the magnetosheath and tended to travel primarily in the direction of the bulk plasma flow (fig. 6–8). Two or three minutes before crossing the shock front, the high-energy particles began arriving from the direction of the solar plasma bulk flow; they continued to do so during the passage through the magnetosheath. After crossing into interplanetary space, the high-energy tail disappeared. Apparently, the high-energy tail was composed of solar plasma particles penetrating through the magnetosheath and eventually swerving to travel in the direction of the bulk flow within the magnetosheath. To attain the fit shown in figure 6–8, a thermal speed of 50/km sec was assumed in the direction of the bulk velocity within the magnetosheath and 70 km/sec perpendicular to this direction for the high-energy tail. Because Pioneer 6 penetrated the magnetosheath during a very quiet period, the above observations are probably characteristic of the normal interaction of the solar plasma with the Earth.

The electron flux was more complex, with three distinct regions being observed. The first region, from 9 to 11.5 Earth radii, was characterized by angularly isotropic fluxes in all four electron channels. The electron energy spectrum indicated that the electrons formed a plasma sheet in

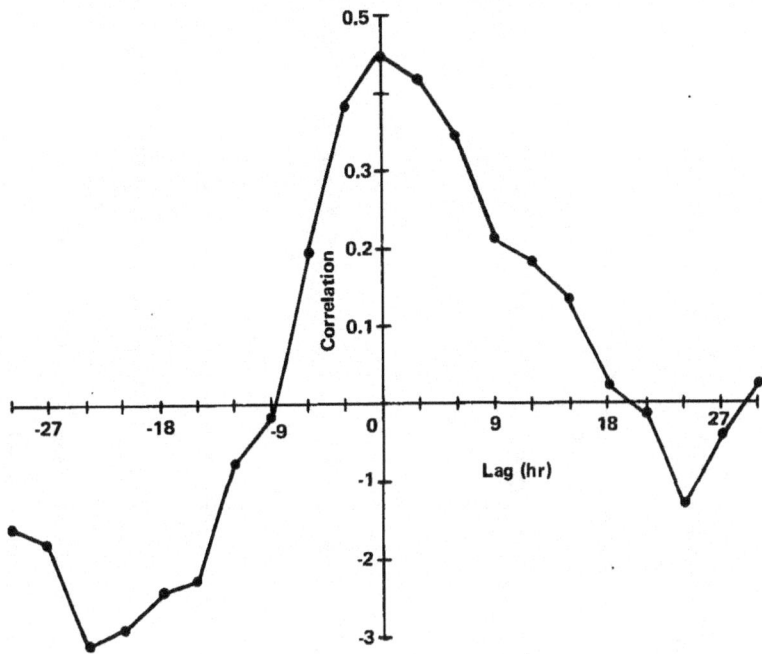

FIGURE 6-6.—Correlation between Pioneer-6 measurements of solar wind speed and angle in the ecliptic plane as a function of lag time. From: ref. 16.

this region. The second region, 1.5-Earth radii thick, was bounded at the outer edge by the magnetopause. The electron distribution in this region could be explained by two models. Using the thermodynamic model presented by Howe, the distribution matched that of a Maxwellian having a pressure of about 300 eV/cm^3, with the temperature parallel to the local magnetic field about twice that perpendicular to the field. In the third region, the magnetosheath itself, the following parameters were typical:
(1) Thermal electron energy—40 eV
(2) Electron speed—2700 km/sec
(3) Electron temperature — 100 000° K

Howe also compared the results of the MIT and Ames plasma probes in this region. MIT velocity measurements were consistently about 20 km/sec higher and well outside the uncertainties of the MIT experiment. There was also clear disagreement in the measurement of the out-of-the-ecliptic flow angle. The density pulse detected by the Ames instrument when the shock was crossed could not be detected by the MIT probe. It should be recalled, however, that these two instruments

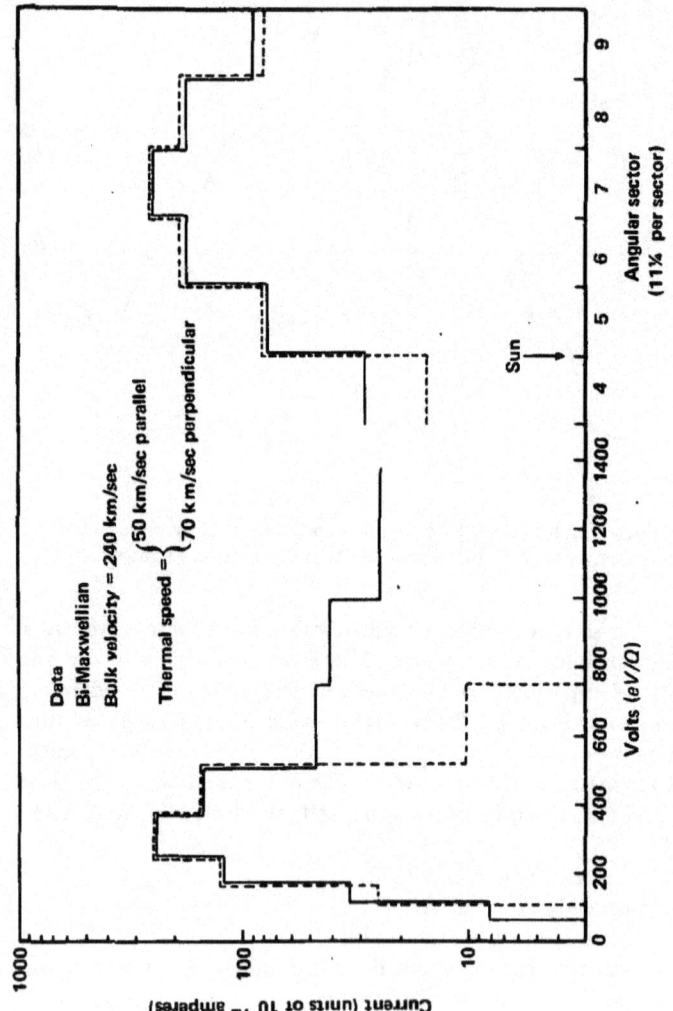

FIGURE 6–7.—Typical magnetosheath energy and angular measurements and the bi-Maxwellian fit. The two thermal speeds are parallel and perpendicular to the bulk velocity, which is within 30° of the magnetic field direction in the magnetosheath. From: ref. 20.

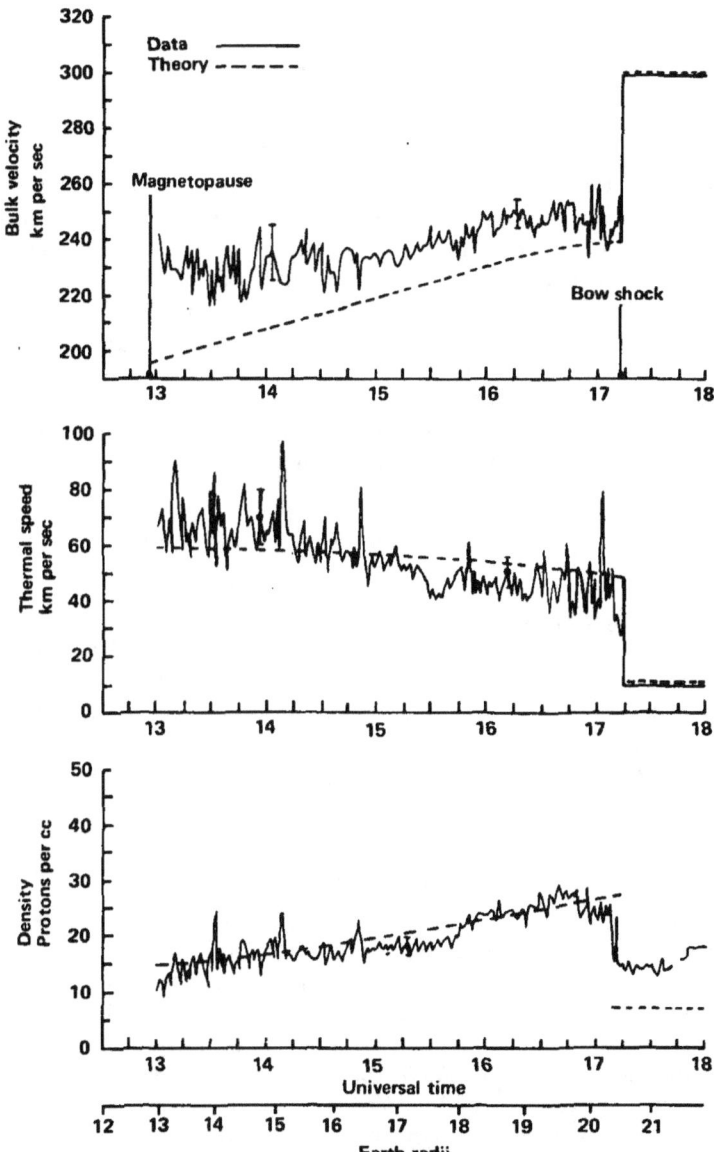

FIGURE 6-8. Pioneer-6 magnetosheath proton observations showing velocity, thermal speed, and number density. From: ref. 20.

are quite different in concept and that one would expect to have to reconcile discrepancies at this stage of their development.

Passage Through the Earth's Tail

The passage of Pioneer 7, an outward bound spacecraft, through the Earth's magnetic tail was recounted in the preceding section. During this passage on August 17 and 18, 1966, data from the MIT probe clearly indicated the existence of a tail and the traversal of the neutral sheet (ref. 21). There was also some evidence of a second neutral sheet near the magnetopause. The data were found to be in general agreement with expectations from a quasistatic model of the geomagnetic tail, based on a balance of particle and field pressures (fig. 6–9). Also shown in figure 6–9 is the apparent correlation of a period of low particle flux with the terrestrial observation of a geomagnetic bay (insert).

THE AMES PLASMA PROBE (ALL PIONEERS)

The Block-I and Block-II plasma probes (called quadrispherical electrostatic analyzers) built by Ames Research Center record the energy spectra of electrons and positive ions in the solar plasma as functions of azimuth and elevation angles (see ch. 5, Vol. II). For a more complete understanding of the interplanetary medium, it is essential to relate plasma probe results to the magnetometer data and, of course, the somewhat different perspectives apparent to the MIT Faraday-cup plasma probe and the TRW Systems electric field detector.

Some Early Results

Like the other Pioneer-6 experimenters, the Ames plasma-probe group presented preliminary results at the 1966 Pioneer-6 Symposium sponsored by the American Geophysical Union (ref. 22). Figures 6–10 and 6–11 show two basic types of data acquired by the Ames plasma probe —energy spectra and angular spectra. The energy spectrum (fig. 6–10) indicates a proton peak at 1350 V, corresponding to a proton velocity of approximately 510 km/sec. The second peak in the curve was due to alpha particles. However, analysis of subsequent data revealed the possible presence of singly ionized helium in the solar wind—the first time this had been detected. In the angular spectra (fig. 6–11), collector 5 consistently recorded higher fluxes than collector 4. The inference was that a net southward convective flow of plasma existed with respect to the plane of the ecliptic. There was also an obvious velocity dependence in ecliptic longitude. It was quite apparent from early Pioneer-6 meas-

SCIENTIFIC RESULTS

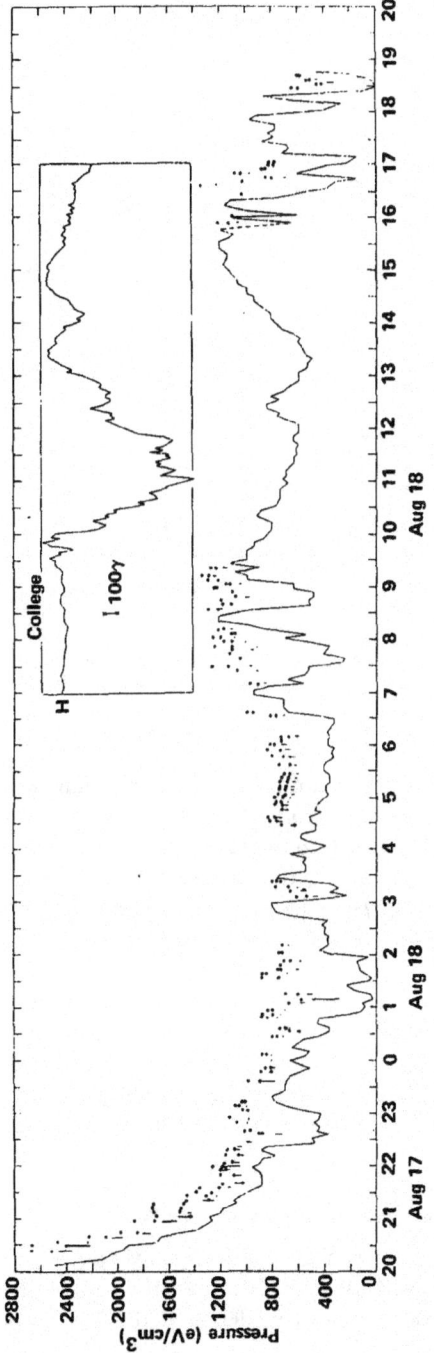

FIGURE 6-9.—Magnetic field pressure (solid line) and total pressure (data points) assuming equal proton and electron pressure. A magnetic bay at College, Alaska, is shown in the inset with the same time scale as the data record. The bay event coincides with a decrease total pressure in the tail. From: ref. 21.

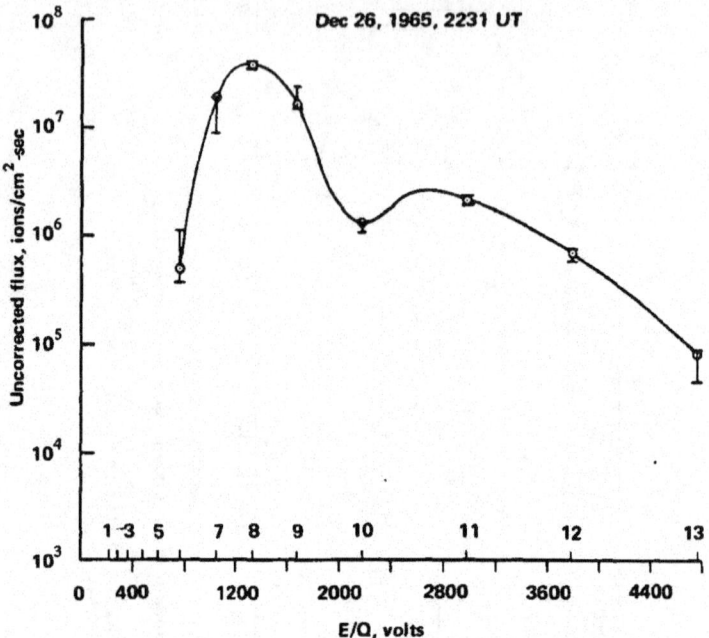

FIGURE 6-10.—Pioneer-6 Ames plasma probe E/Q spectrum, Dec. 26, 1965, 2231 UT, showing the hydrogen peak at approximately 1350 V with the helium peak estimated at 2700 V. From: ref. 22.

urements that the common assumptions of solar radial flow of plasma and thermal isotropy were not valid.

The early data also revealed an average solar wind electron temperature of about 100 000° K during quiet times when the solar wind was blowing at about 290 km/sec, with a maximum ion temperature of 50 000° K. Interplanetary electrons always seem hotter than ions during quiet periods.

As Pioneer 6 passed through the Earth's magnetopause, the Ames plasma probe measured the temperature of solar electrons in the bow shock at 500 000° K. Here, ion temperatures were about the same as electron temperatures, but, in contrast, the ions did not cool off downstream from the Earth. The ions also exhibited other non-thermal characteristics.

Observations of the Earth's Wake

Pioneers 7 and 8 were outward-bound missions and, as illustrated in figure 6-12, swept through the Earth's tail early in their flights. Instruments on both spacecraft detected evidence of the Earth's tail or wake

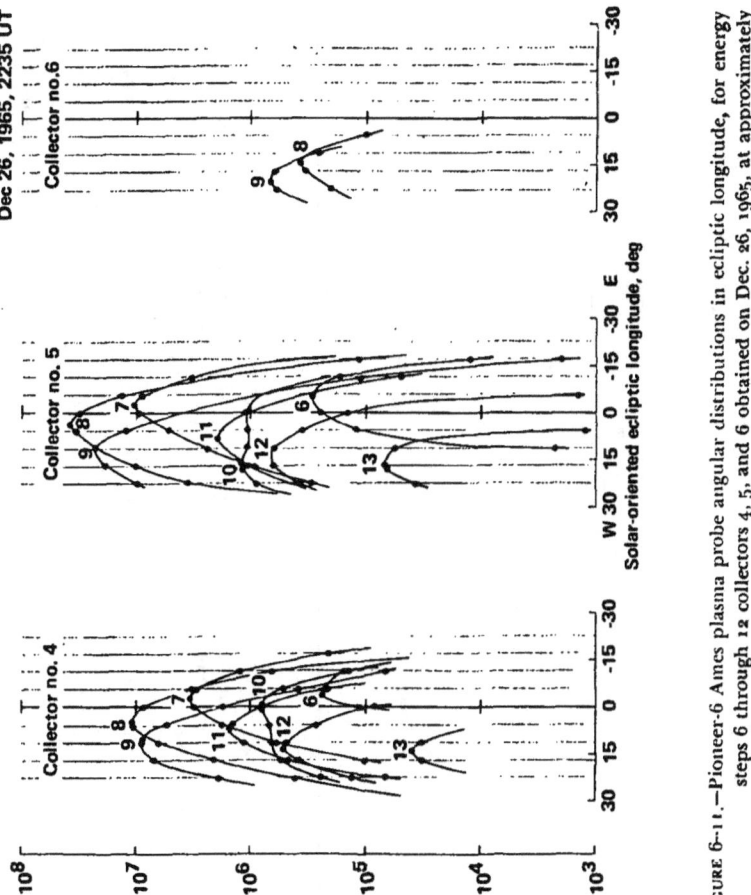

FIGURE 6-11.—Pioneer-6 Ames plasma probe angular distributions in ecliptic longitude, for energy steps 6 through 12 collectors 4, 5, and 6 obtained on Dec. 26, 1965, at approximately 2235 UT. From: ref. 22.

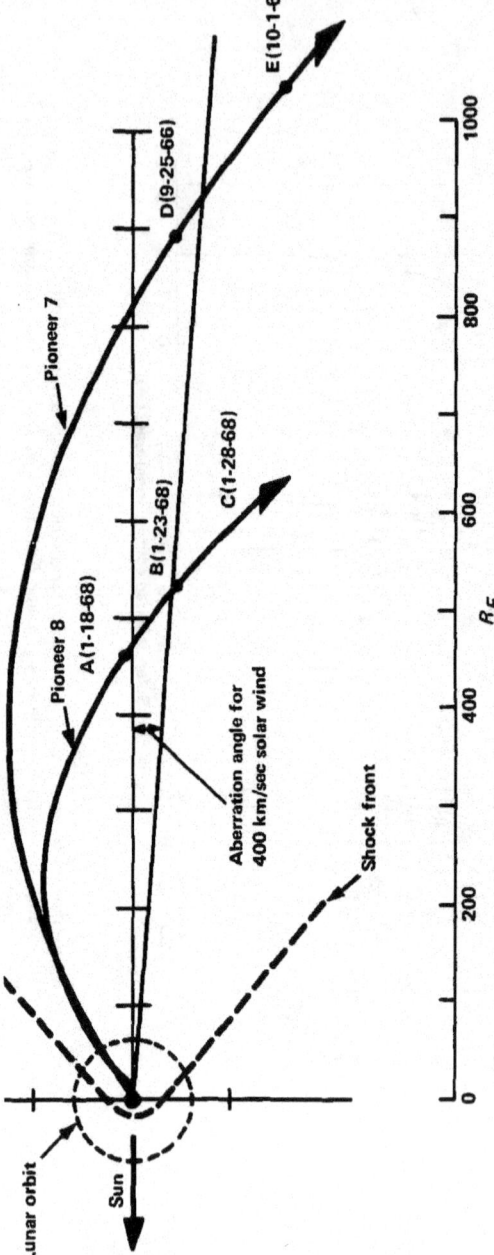

FIGURE 6-12.—Pioneers-7 and -8 trajectories; ecliptic projection. Pioneer 7, launched Aug. 17, 1966, went through the expected region of the geomagnetic tail at 1000 R_E in Sept. 1966. Pioneer 8, launched Dec. 13, 1967, went through the expected region of the geomagnetic tail at 500 R_E in Jan. 1968. The Pioneer-8 spacecraft coordinates at locations A, B, and C are, respectively, 460 R_E, 525 R_E, and 590 R_E geocentric from the Earth; 8.5 R_E, 10.5 R_E, and 12.3 R_E above the ecliptic at an ecliptic projection of the spacecraft-Earth-Sun angle of 180°, 185.5° and 190.5°. The Pioneer-7 spacecraft locations at D and E are, respectively, 887 R_E and 1059 R_E geocentric from the Earth, 25.4 R_E and 28.7 R_E above the ecliptic at an ecliptic projection of the spacecraft-Earth-Sun angle of 183° and 189°. From: ref. 23.

with their magnetometers and plasma probes. The Ames plasma probes detected the wakes at about 1000 and 500 Earth radii for Pioneers 7 and 8, respectively (ref. 23). In each case the normally quiescent plasma ion energy spectra were interrupted by the abrupt changes in the magnitude and curve shape that one would expect near the tail boundaries.

In figure 6-13, the typical quiescent ion spectrum is compared with that measured in the Earth's wake by Pioneer 7. The peaks of the "disturbed" spectra are usually one or two orders of magnitude less than those of quiescent spectra. Further, there is often a different kind of double peak (indicated by the dashed lines) which infers that the

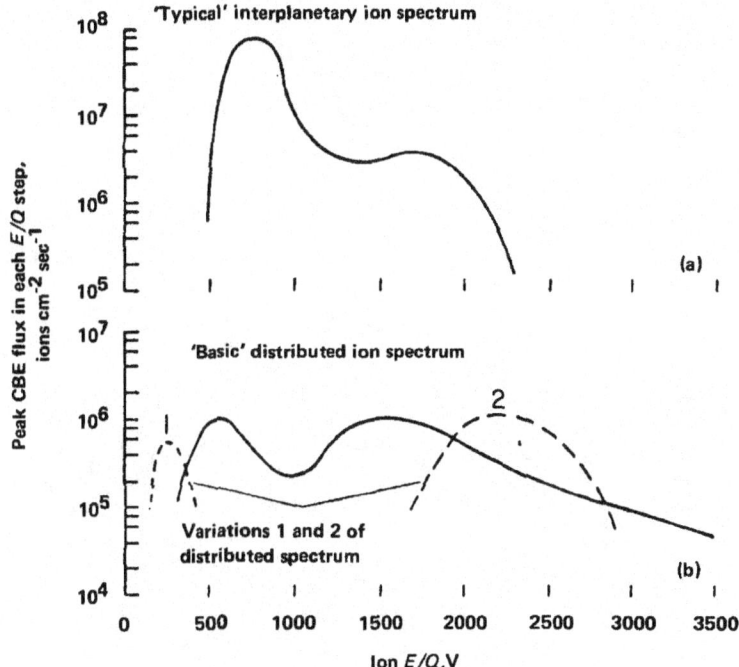

FIGURE 6-13.—Pioneer-7 ion spectrum. (a) 'Typical' interplanetary ion spectrum. The peak CBE flux of the curve (the H^+ peak) is 10^7–10^8 ions cm^{-2}sec^{-1}. The energy per unit charge is 850 V. The second peak at 1700 V (2×850 V) is the He^{++} peak. (b) 'Basic' disturbed ion spectrum and two variations often observed in the geomagnetospheric wake. The peak flux is $< \sim 10^6$ ions cm^{-2}sec^{-1}. In this case the first peak of the curve (the H^+ peak) occurs at ~ 500 V and the second at ~ 1500 V. The second peak is interpreted as a high energy tail of the proton energy distribution. Analysis of successive ion spectra show that the higher energy distribution often grows at the expense of the lower energy distribution. At times only part of the distribution is seen as indicated by the two variations. From: ref. 23.

high energy tail of the proton distribution may grow at the expense of the low energy protons in the vicinity of the Earth's wake. The disturbed nature of plasma conditions in this region is quite apparent in figure 6–14 where the energy spectra are seen to vary markedly with time. During some periods, for example, no flux could be detected at all; at other times, the spectra were either normal or disturbed, indicating a mixing that seems reasonable near the turbulent boundaries of a magnetic tail.

The Ames investigators felt, on the basis of their data, that the following interpretations were possible:

(1) The observations could represent a turbulent downstream wake if the Earth's magnetosphere closed between 80 and 500 Earth radii.

(2) If the solar wind diffuses into the magnetic tail, the plasma probe measurements could be due to the tail "flapping" past the spacecraft.

(3) The tail might have a filamentary structure at these distances (500 and 1000 Earth radii), and the disturbed data could arise at filament boundaries.

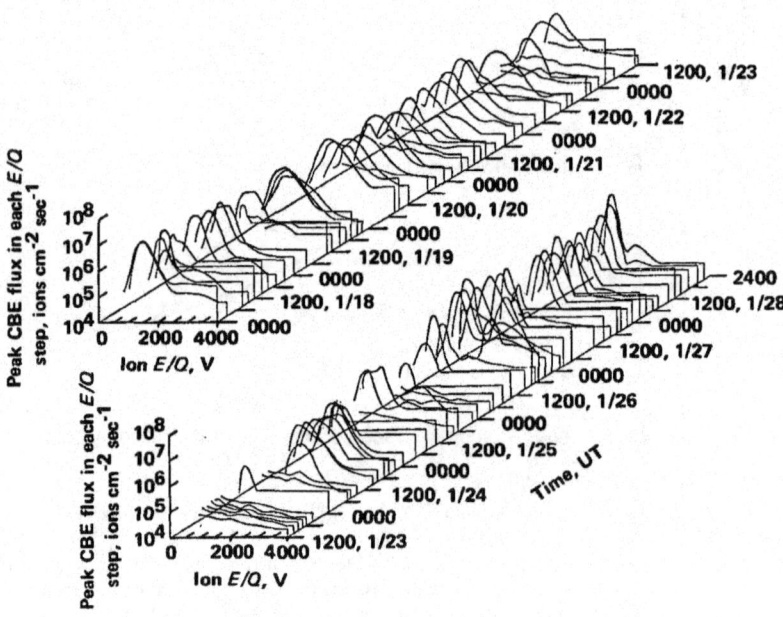

FIGURE 6–14.—Pioneer 8 ion spectra, Jan. 18–28, 1968. Eleven days of spectra (approximately 2 hr apart) are plotted. The changing nature of the measurable plasma characteristics are illustrated. Real time teletype data gaps are indicated by not plotting any spectrum (e.g., 2000-2400 UT on Jan. 23, 1968), while an absence of measurable plasma is indicated by plotting a low-lying noise spectrum (e.g., 1200-1800 UT on Jan. 23, 1968). From: ref. 23.

(4) Possibly the tail might have disintegrated into "bundles" at these distances.

(5) If magnetic merging occurred, subsequent acceleration of pinched-off gas may have caused the disturbed conditions measured.

Plasma Instabilities

Prior to Pioneer 6, few spacecraft were capable of making detailed measurements of the solar wind. Consequently, the collisionless interplanetary plasma was treated as a single magnetofluid. However, the Ames plasma probes have revealed that the solar proton distribution is definitely anisotropic, with the temperature parallel to the local magnetic field being larger than that perpendicular to the local magnetic field (ref. 24). From these data and basic theory, it can be shown that the anisotropy can be produced by the approximate conservation of magnetic moment and thermal energy as the collisionless solar plasma flows outward and the imbedded magnetic field weakens. The positive ion distributions measured were also unstable with respect to the generation of low-frquency whistlers. The conclusion was that a generalized form of "firehose" instability must occur with the growth of whistlers near the ion cyclotron frequency.

THE CHICAGO COSMIC RAY EXPERIMENT (PIONEERS 6 AND 7)

The Chicago cosmic-ray telescope on the Block-I Pioneers provided the opportunity for scientists to investigate the direction of arrival of cosmic-ray particles near the plane of the ecliptic. The experiment also had a short enough time resolution so that rapid fluctuations in cosmic-ray intensity could be recorded. The first test case came shortly after the launch of Pioneer 6 when solar-flare protons were detected on December 30, 1965. These early results—some of them unexpected—were reported at the Pioneer-6 Symposium by the Chicago group (ref. 25).

Anisotropy of Solar-Flare Proton Flux

The solar flare that erupted about 2 weeks after the launch of Pioneer 6 was given an importance rating of 2. The effects were noted for almost a week, as indicated in figure 6–15. Interplanetary conditions during most of this period were remarkably free of solar-flare blast effects capable of modulating the galactic cosmic-ray flux. Solar protons in the energy range 13 to 70 MeV first arrived at the spacecraft at about 0300 UT, December 30, 1965, with lower energy particles arriving later. The anisotropy of these protons was striking (fig. 6–16), with the average direction of particle flow about halfway between the Sun line and the

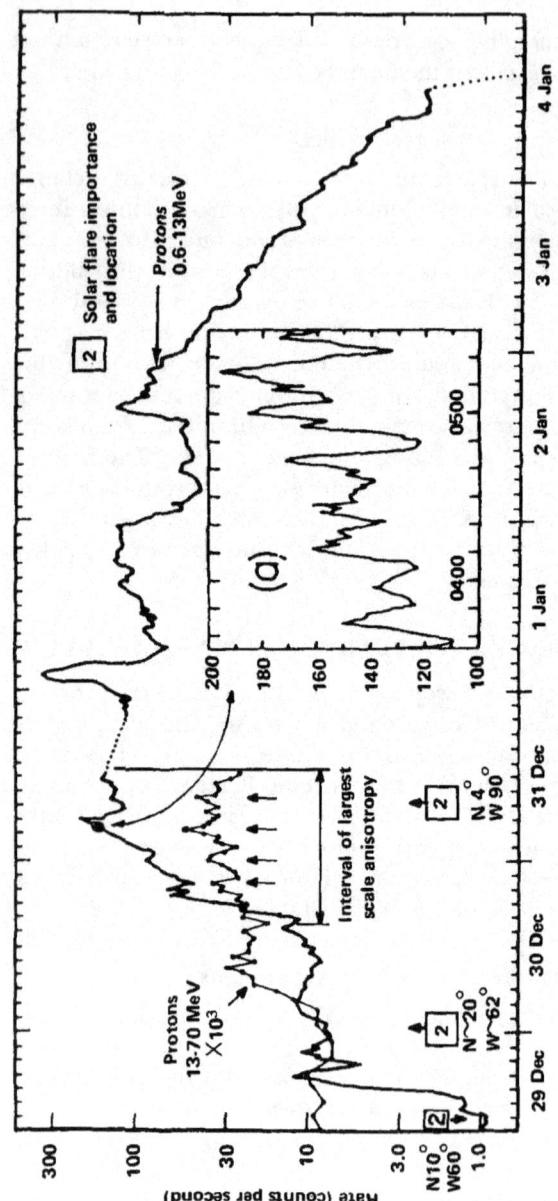

FIGURE 6–15.—The intensity-time distribution of protons of 0.6–13-MeV energy and protons 13–70-MeV energy. Anisotropies were observed for a period of approximately 2 days after the flare of Dec. 30, 1965. The arrows refer to quasi-periodic bursts of period ∼4 hr. Insert (a) is an expansion of the region shown within the circle. Data points are ∼56 sec apart. Note the quasi-periodic oscillations of ∼15 min. From: ref. 25.

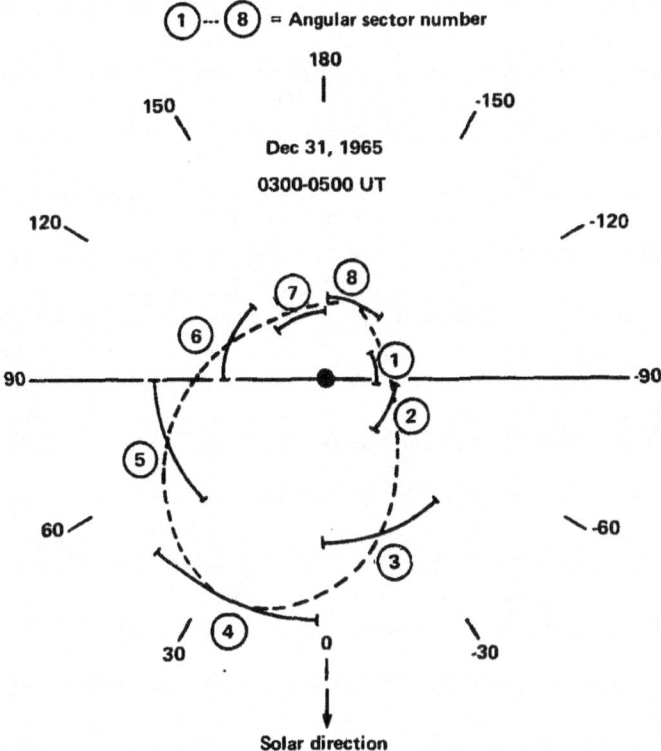

FIGURE 6-16.—Anisotropy diagram for proton flux from flare recorded Dec. 30, 1965. From: ref. 25.

angle one would expect if the particles travelled along the "water-sprinkler" spiral lines. However, the detailed data reveal a more complex situation:

(1) The direction of the peak amplitude was highly variable, changing direction by as much as 90° within 10 min.

(2) Relative to the intensities in other directions, the peak intensity varied rapidly.

(3) Occasionally, the angular distribution was strongly peaked within a 45° sector.

(4) Rarely, two intensity peaks 180° apart were noted.

The strong collimation of solar protons with energies greater than 13 MeV infers that there are few irregularities in the propagation path from the Sun that could scatter the protons. However, the rapid changes in direction of the peak flux vector supports the conclusion from Goddard magnetometer and GRCSW cosmic-ray anisotropy data that there are many short-term, rather localized changes in the Earth's magnetic field. (See discussion of the possible filamentary character of interplanetary space under these other experiments.) The double intensity peak noted on occasion implies that some back-scattering does occur out beyond the spacecraft's orbit.

Rapid Intensity Variations

During part of the solar-flare event (about 7 hr), the proton measurements at energies of the order of 600 keV displayed large-scale, quasi-periodic bursts with periods of about 900 sec and characteristic rise and fall times of roughly 100 to 200 sec. In addition, quasi-periodic fluctuations were noted in the 13 to 70 MeV energy range with periods of 3.5 to 4 hr. It is possible that these fluctuations indicate the existence of Alfven waves in the inner solar system.

Sector Structure of Interplanetary Space

Corotation effects were noted early in flight by the Chicago instrument, supporting the joint observations of several other Pioneer 6 instruments and similar instruments on spacecraft elsewhere in the solar system. For example, proton intensity structures detected at Pioneer-6 were noted some 2 hr later at the IMP-3 (Explorer 28) Earth satellite.

Proton flux increases over the period from December 1965 through September 1966 have been unambiguously associated with specific solar flares (ref. 26). Enhanced solar proton fluxes in the energy range 0.6 to 13 MeV have been recorded from specific active regions from ranges as great as 180° in longitude. The enhanced fluxes were characterized by definite onsets when their associated active centers reached points 60° to 70° east of the central solar meridian. Cutoffs occurred 100° to 130° west. Coupled with the detection of associated modulations of the galactic cosmic-ray flux, these observations again point to the existence of corotating magnetic regions associated with the active centers on the Sun (fig. 6-17). Observations seem to show that solar-flare protons propagate along the spiral interplanetary field from the Sun's western hemisphere. Present evidence supports the view that the solar protons arise from processes continually occurring in the solar active centers.

The protons from flares located at ranges over 60° in longitude seem to propagate rapidly through the corona into the interplanetary magnetic field. The short transit and rise times cannot be explained by isotropic diffusion across coronal magnetic fields. To account for these observations, a new quasi-stationary model was suggested in which some field lines rooted in or near the active center are spread out in the corona over a range of about 100° to 180° in longitude and then extended into interplanetary space by the solar wind.

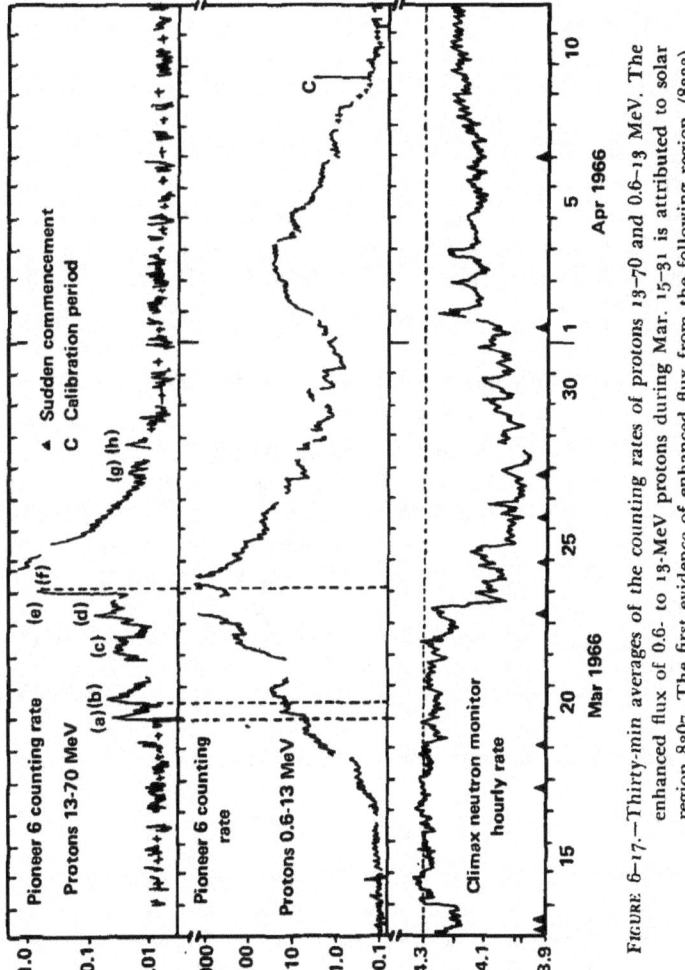

FIGURE 6-17.—Thirty-min averages of the counting rates of protons 13-70 and 0.6-13 MeV. The enhanced flux of 0.6- to 13-MeV protons during Mar. 15-31 is attributed to solar region 8207. The first evidence of enhanced flux from the following region (8223) appears on Mar. 31. A magnetic sector boundary occurs on Mar. 31. Note at this time the abrupt change in the level modulation at the Climax neutron monitor. (a)-(h) denote discrete flare and shock-wave events seen at 13-70 MeV. From: ref. 26.

Differential Energy Spectra

By correlating measurements from Chicago cosmic-ray telescopes aboard Pioneer 7 and OGOs 1 and 3, the Chicago group has shown that protons and helium nuclei in the 1 to 20 MeV/nucleon range are present during the so-called "quiet times" often observed in interplanetary space from 1964 to 1966 (ref. 27). Further, the observed helium nuclei flux was shown to increase from 1964 to 1965 and then decrease from 1965 to 1966—in accord with the observed variation of galactic helium nuclei by terrestrial detectors. In contrast, the proton flux detected kept increasing during these 3 yr.

These results infer that most of the particles observed by the spacecraft at the time of solar minimum are of galactic origin. As the new solar cycle began, solar particles (mostly protons) began to enhance the galactic proton flux. The H/He ratio rose from 2 in 1965 to about 10 in 1966.

Relativistic Electrons in the Geomagnetic Tail

NASA's scientific satellites have established that a neutral sheet (where the Earth's magnetic field is essentially nil) exists within the Earth's tail between about 11 and 80 Earth radii. Satellites have also detected high energy electrons streaming along this sheet. During its passage through the geomagnetic tail in August 1966, Pioneer 7 observed relativistic electrons confined within this neutral sheet at about 19 and 38 Earth radii (ref. 28). The two high-energy-electron peaks (>400 MeV), shown in figure 6–18, were coincident with Pioneer-7 passages across the neutral sheet. The relativistic electron fluxes did not extend outside the neutral sheet, and the evidence points to acceleration of the electrons by the split magnetic field within the sheet. The origin of this unique feature in nature is still controversial.

THE GRCSW COSMIC RAY EXPERIMENT (ALL PIONEERS)

The primary mission of the GRCSW experiment was the measurement of anisotropy in the distribution of cosmic rays within the solar system but still far enough away from the Earth to avoid its perturbing magnetic field. The construction of a theoretical model describing how cosmic rays are propagated through the solar system depends upon the accurate measurement of cosmic rays with energies less than 1000 MeV. Because the weaker cosmic rays, especially those originating on the Sun, are affected by the solar magnetic field and the plasma in which it is imbedded, the GRCSW data must be examined in conjunction with the results of the Pioneer plasma and magnetometer experiments. Some

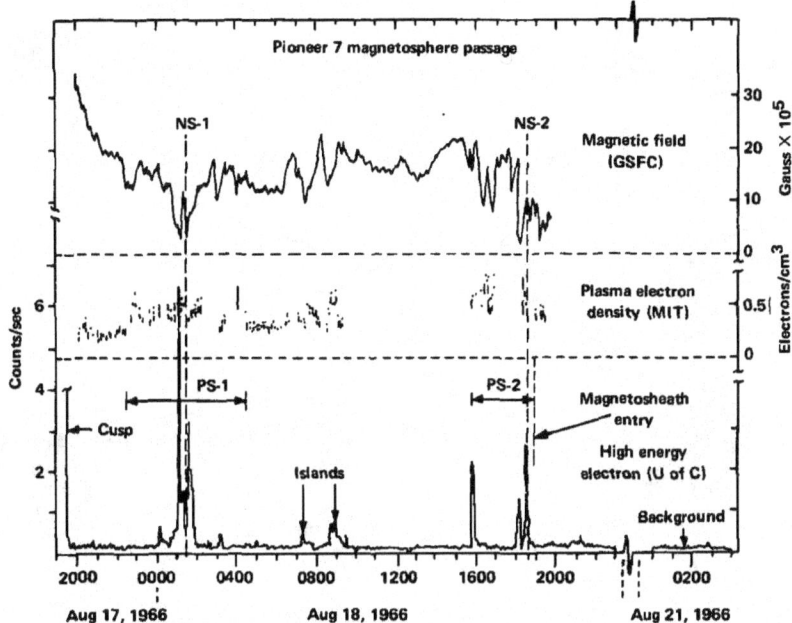

FIGURE 6-18.—The intensity of electrons >400 keV as a function of time as the spacecraft emerges from the cusp region until passage into the magnetosheath (ref. 28). A typical cosmic ray background flux is shown on the right for Aug. 21. For comparison the magnetic field strength and electron concentration reported by Lazarus et al. (ref. 21) are shown on the same scale. NS-1 = neutral sheet passage no. 1.

of these interrelationships have already been discussed in the preceding sections.

General Structure of Interplanetary Space

The early data from Pioneer 6 revealed that the low-energy cosmic-ray flux (13 MeV/nucleon) exhibited considerable anisotropy. It was this anisotropy which, when combined with Goddard magnetometer data, led to the original concept of the filamentary structure described earlier in connection with the Goddard magnetometer (ref. 29). The close correlation of these two sets of data can be seen in figure 6-3. According to this model, cosmic rays of low energy are forced to flow inside these twisting tubes, as shown in figure 6-19. The changing cosmic-ray anisotropies were ascribed to the experiment's sampling of the fluxes in the various tubes as they swept past the spacecraft. The tubes themselves were estimated to be between 0.5 and 4 million km in diameter. As the

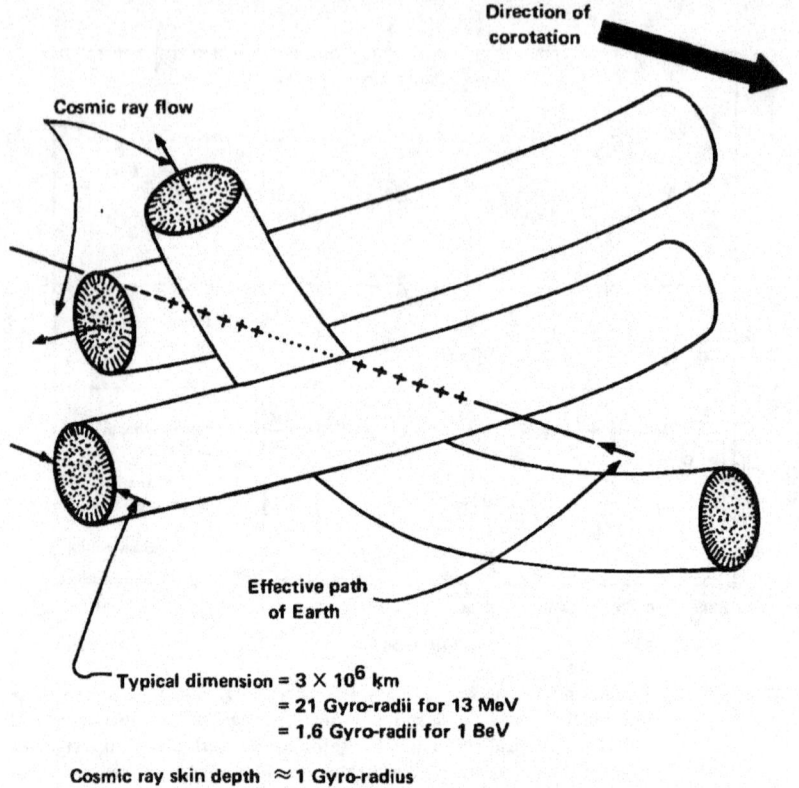

FIGURE 6-19.—A simplified model of the filamentary structure of the interplanetary magnetic field. Each filament can be thought of as a bundle of tubes of force. The cosmic rays of low energy are constrained to travel along the filament by the magnetic field. From: Bartley; et al.: J. Geophys. Res., vol. 71, Jul. 1, 1966, p. 3301.

reader will recall, later magnetometer data suggested that the filament model should be replaced by a "discontinuity" model.

The extent of the anisotropy of low-energy solar protons during early flight was striking. Since scattering normally reduces anisotropy, these results imply that little scattering transpired since the cosmic rays were injected into the interplanetary field near the Sun. In contrast, the anisotropy of relativistic cosmic rays is known to be obliterated quickly.

In 1967, Rao, McCracken, and Bartley (ref. 30) summarized these analyses of Pioneer anisotropy data collected during 1965 and 1966 for the period when solar flare effects were not seen. Considering only cosmic rays in the vicinity of 10 MeV/nucleon, their conclusions were:

(1) The 10 MeV/nucleon cosmic rays possessed a density gradient directed toward the Sun; i.e., density increases sunward, as expected.

(2) These low-energy cosmic rays are predominantly of solar origin even during the sunspot minimum.

(3) The density gradient frequently reverses in the range $10 < E < 1000$ MeV.

(4) Cosmic radiation between 10 and 10^5 MeV corotates with the Sun.

Studies of the large-scale, steady-state structure of interplanetary space have also been made by comparing Pioneer data with those from other spacecraft (ref. 31). Comparison of Pioneer cosmic-ray telemetry with comparable data from the IMP 3 (Explorer 28) Geiger counter showed close agreement when the spacecraft were lined up with the sun. When separated by 50° in azimuth (in late 1966) the variations in cosmic-ray flux appeared to be due mainly to galactic cosmic rays. Balasubrahmanyan and his colleagues concluded that there exist numerous, long-lived regions of modulated cosmic-ray flux following the general spiral configuration of the interplanetary magnetic field as is corotates with the Sun.

A Closer Look at the Anisotropy-Magnetic Field Relationship

The early paper of McCracken and Ness (ref. 29), which introduced the filament concept, was modified by another joint paper in 1968 (ref. 32). The main thrust of this paper was that the observed anisotropies of low-energy cosmic rays could be divided into two groups:

(1) Equilibrium anisotropies which are most evident toward the end of a solar-flare event. The maximum cosmic-ray flux is always directed away from the Sun (fig. 6–20), and the anisotropy amplitude is low (5 to 15 percent). Perhaps of most significance is the fact that the anisotropies are not dependent upon the detailed nature of the interplanetary magnetic field.

(2) Nonequilibrium anisotropies which change direction in time and have amplitudes between 20 and 50 percent. These anisotropies are aligned—parallel or antiparallel—to the magnetic field.

These observations were interpreted as possible evidence of complex loops in the magnetic field.

Cosmic-Ray Propagation Processes

The GRCSW group published two general papers relating Pioneer cosmic-ray data to cosmic-ray flare effects and energetic-storm-particle events (refs. 33 and 34). The data used came from Pioneers 6 and 7 and covered 29 solar flares occurring between December 16, 1965, and October 31, 1966. Some of the more important conclusions expressed in this paper were:

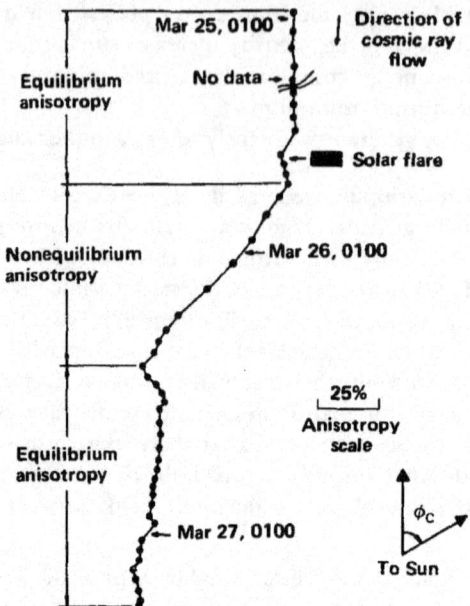

FIGURE 6–20.—The difference between the equilibrium and nonequilibrium classes of cosmic-ray anisotropy. The amplitudes and azimuths of the mean anisotropy for each hour are plotted as a vector addition diagram. Note definition of ϕ_c. From: ref. 32.

(1) Solar cosmic rays are normally extremely anisotropic with the direction of maximum flux aligned parallel to the magnetic field vector during the first part of the solar event.

(2) During the late portion of the flare, the cosmic rays are in diffusive equilibrium.

(3) Under some circumstances, the propagation of cosmic rays from the Sun to Earth is completely dominated by a "bulk motion" propagation mode. Here, the cosmic rays do not reach the spacecraft until the magnetic regime into which they were injected engulfs the Earth.

(4) In two cases, the anisotropy and cosmic-ray times of flight infer diffusion of the cosmic rays to a point on the western portion of the solar disk before injection into the magnetic field.

(5) Simultaneous observation by both Pioneers when separated by 54° of azimuth indicate density gradients of about two orders of magnitude per 60° sector during the initial stages of a solar flare.

(6) A study of cosmic-ray scattering within the solar system indicates a mean free path of about 1.0 AU for large-angle scattering.

The second paper dealt with the energetic-storm-particle event which was defined as the very marked enhancement of cosmic rays in the 1 to

10 MeV range near the onset of a strong terrestrial magnetic storm. Data relating to seven such events were extracted from Pioneer-6 and Pioneer-7 telemetry. The data indicated a near one-to-one correspondence between the energetic-storm-particle events and the beginning of a Forbush decrease (fig. 6-21). It was shown further that the bulk of the energetic-storm-particles are apparently not trapped in the magnetic regime associated with the Forbush decrease. The Pioneer cosmic-ray data tend to support the Parker "blast wave" model in which the charged particles are accelerated by the magnetic field within the shock front. Further discussion can be found in reference 35.

The GRCSW group also compared the characteristics of corotating the flare-induced Forbush decreases as derived from cosmic-ray data obtained from Pioneers 6 and 7 (ref. 36). The results of this investigation are summarized in table 6-1.

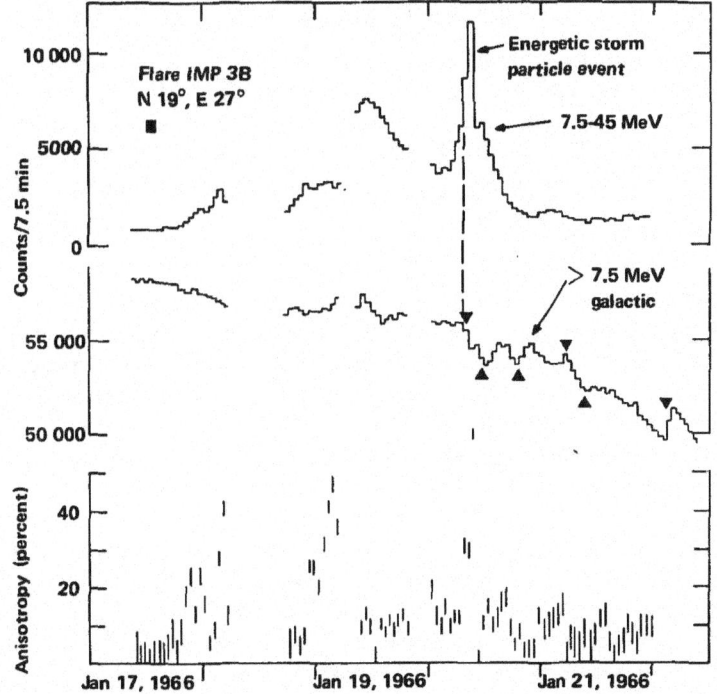

FIGURE 6-21—Temporal variations of the cosmic-ray counting rates and the cosmic-ray anisotropy amplitude during the period Jan. 17–21, 1966. Note that the major portion of the energetic storm-particle event occurred after the onset of the small Forbush decrease. The solid wedges refer to times at which major changes were observed to occur in the anisotropic nature of the cosmic radiation. From: ref. 34.

TABLE 6–1.—*Comparison of the Properties of Corotating and Flare-Initiated Forbush Decreases*[a,b]

Corotating Forbush decrease	Flare-initiated Forbush decrease
Not accompanied by solar-generated cosmic rays	Accompanied by solar cosmic rays and an energetic storm particle event
Onset time difference due to corotation	Probably simultaneous onset up to ∼100° off the axis of the Forbush decrease
No amplitude dependence over ∼60° of solar azimuth	Amplitude varies by a factor of ∼4.0 over ∼60° of solar azimuth

[a] Adapted from reference 36.
[b] The energy dependence of both classes of events is esentially the same.

Galactic Alpha-Particle Flux

Some limited studies of the galactic alpha-particle flux measured by the Pioneer-6 GRCSW cosmic-ray experiment have been reported (ref. 37). An examination of the time dependence of alpha particles in the 124- to 304-MeV range shows that these particles exhibit the same recurrent Forbush decreases previously observed in the galactic proton flux.

Studies of Specific Solar-Flare Events

Several solar-flare events have been examined in detail in the light of GRCSW cosmic-ray data and readings taken at several ground stations. By way of illustration, the results of the studies of the January 28, 1967, and March 30, 1969, events are summarized below (ref. 38). The salient features of the first event were:

(1) The probable location of the responsible solar flare was about 60° beyond the west limb of the Sun.

(2) Low-energy particles (< 100 MeV) recorded by the Pioneers and the high-energy particles (> 500 MeV) detected at Earth arrived after diffusion across the interplanetary magnetic field. Both groups of particles displayed remarkable isotropy.

(3) The flux that would be observed by a detector ideally located in azimuth would be greater than 2000 particles cm.$^{-2}$ sec.$^{-1}$ sr.$^{-1}$ above 7.5 MeV.

(4) Pioneer observations indicated low-energy injection commencing several hours before the high-energy main event.

Reiff has written a running account of the March 30, 1969, event (ref. 39). Solar activity was high during most of March; several active regions capable of generating solar flares were under surveillance. On March 30, after the most active of these regions had rotated behind the west limb of the Sun, terrestrial radio telescopes recorded the largest 10-cm burst from the corona in scientific history. Within about 2 hr, the cosmic-ray instrumentation on Pioneers 6, 8, and 9 noted a sharp increase in low- and high-energy protons. About a day later, Pioneer 7 recorded the same increase in flux. Apparently a large solar flare had occurred on the other side of the Sun. The fluxes recorded by Pioneers 8 and 9 as well as those on Earth satellites subsided within a few days; by April 5, Pioneer-6 and -7 data followed suit.

By April 10, the active region of the Sun which produced these effects was only 20° behind the east limb. Once again a flare erupted. Within a half hour, cosmic-ray intensities at Pioneers 6 and 7 jumped more than an order of magnitude. Terrestrial instruments and those on Pioneers 8 and 9 showed little change. The relative locations of the Earth and the spacecraft are indicated on figure 6–22. Evidently the flare-generated radiation first engulfed Pioneers 6 and 7. Two days later, Pioneers 8 and 9 and terrestrial instruments were recording increases while levels at the other Pioneers dropped to near-normal levels. The motion of the active region and solar rotation had combined to turn the spray of radiation more than 90°.

In 1971, a key paper was published by McCracken and his colleagues describing the decay phase of typical solar flares (ref. 40). Some of the important conclusions from this paper follow:

(1) At times less than 4 days after the injection of a solar flare, the anisotropy at 10 MeV tends to be directed radially away from the Sun. After 4 days, this anisotropy is directed 45° east of the spacecraft-Sun line. This situation implies the dominance of convection over diffusion in the escape of solar cosmic rays late in flare life.

(2) A positive radial cosmic-ray density gradient exists at late times (more than 4 days) near the Earth's orbit. This drives a diffusive current along the interplanetary magnetic lines toward the Sun.

(3) The observed temporal variation of cosmic-ray flux can be ascribed to (a) convective removal of the cosmic radiation, and (b) the corotation of the cosmic-ray population.

(4) The observed rate of change of cosmic-ray flux is critically dependent upon the local value of the gradient in heliocentric longitude for energies less than 10 MeV.

(5) Cosmic-ray spectra indicate that the influence of the longitude gradient upon the observed temporal decay increases toward lower energies.

(6) Late in the solar flare, the spectral exponent near 10 MeV is dependent upon the longitude of the observer relative to the centroid of the cosmic-ray population injected by the flare.

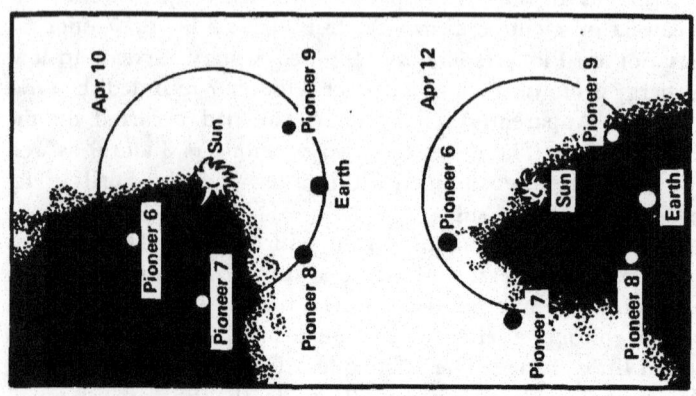
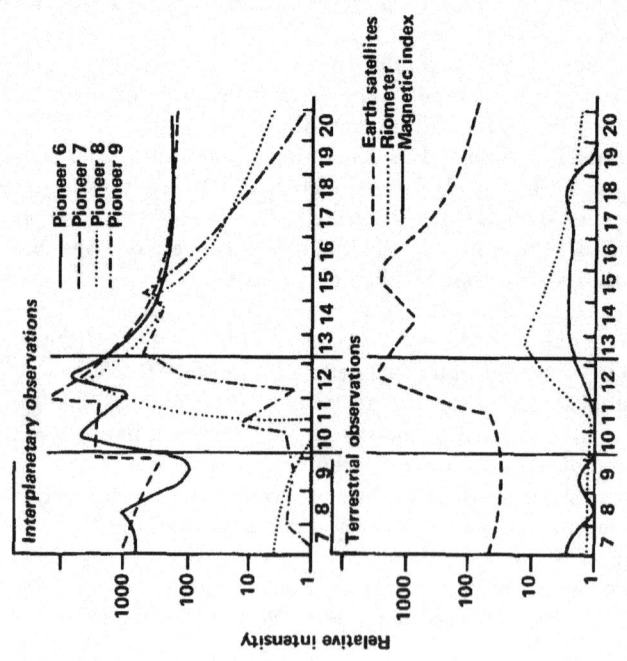

FIGURE 6-22.—Pioneer locations and data summaries for the solar proton event beginning on Apr. 10, 1969. From: ref. 39.

THE MINNESOTA COSMIC RAY EXPERIMENT
(PIONEERS 8 AND 9)

The Minnesota cosmic-ray telescopes replaced the Chicago instruments on the Block-I Pioneer flights. The energy range of the Minnesota instrument was considerably higher (4 MeV/nucleon to over 2 BeV/nucleon). The research results published to date are primarily concerned with galactic cosmic rays rather than the lower-energy particles originating on the Sun, although papers on solar cosmic rays are in preparation.

Composition of Galactic Cosmic Rays

Although the so-called "M (medium) nuclei," carbon, nitrogen, and oxygen are the most abundant nuclei in cosmic rays except for hydrogen and helium, their relative abundances have been in question until recently. New measurements of cosmic-ray nitrogen from balloons and Pioneer 8 have provided better estimates (ref. 41). The energy spectrum of nitrogen was found to be identical with those of the other M nuclei over the range of 100 MeV to over 22 BeV/nucleon. The ratio of nitrogen nuclei to all M nuclei was found to be about 0.125, constant to within 10 percent over the above energy range (fig. 6–23). Assuming that some of the nitrogen in the cosmic-ray flux originates in fragmentation reactions with interstellar matter and knowing the proper cross sections, one can compute a "source" N/M ratio less than about 0.03. However, the solar atmospheric value for the N/M ratio is about 0.10—a disturbingly higher value. The implication is that galactic and solar cosmic rays may originate in fundamentally different processes.

The Pioneer-8 instrument also identified and measured fluorine nuclei in the galactic cosmic rays (ref. 42). The fluorine abundance was 1 to 2 percent that of oxygen for energies above 500 MeV/nucleon. These data on fluorine are consistent with the hypothesis that the fluorine is created by the fragmentation of heavier nuclei as they traverse roughly 4 g/cm^2 of hydrogen in their flights through the galaxy.

In a later paper, the Pioneer-8 data were used to estimate the chemical composition and energy spectra of cosmic rays with atomic numbers from 3 to 30 (ref. 43). Briefly, the results were as follows. The ratio of light to medium elements (L/M ratio) was 0.25 ± 0.02 and was constant with energy over the range of 100 MeV/nucleon to over 22 BeV/nucleon. No significant variations in the individual Li/M, Be/M, and B/M ratios were observed as a function of energy. These ratios were 0.086, 0.037, and 0.150, respectively (fig. 6–24). However, the Be/(Li+B) ratio was considerably less than that predicted from known fragmentation parameters, suggesting that some Be^7 had decayed in flight. The chemical composition of the heavier cosmic rays was roughly what one

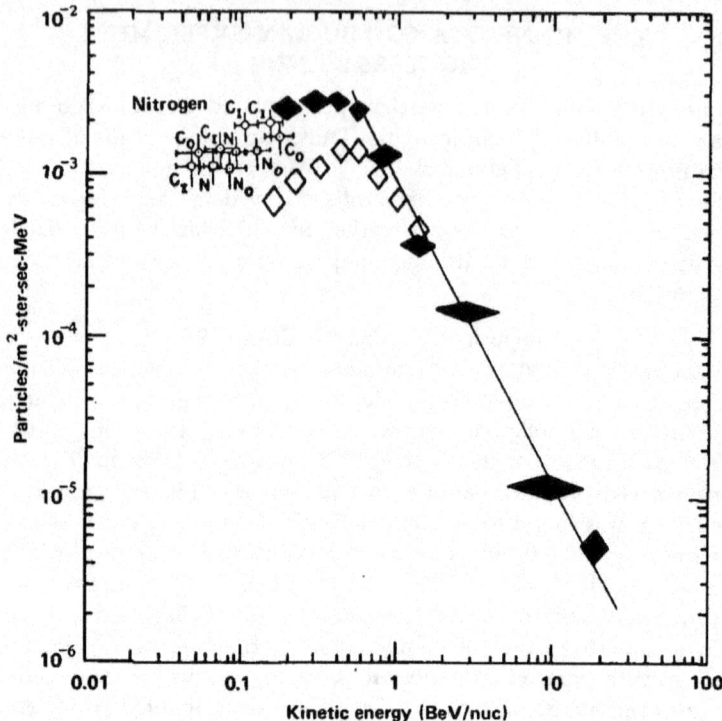

FIGURE 6-23.—Differential spectra of nitrogen nuclei measured by Pioneer 8 in 1968 (open diamonds) and from balloons in 1966 (solid diamonds). The low-energy points are from several satellites. From: ref. 41.

would expect if they originated from the fragmentation of iron in galactic space (table 6-2).

Primary Electrons in the 0.2-MeV to 15-BeV Range

The Pioneer-8 cosmic-ray telescope measured primary electrons at the extreme low end of the energy spectrum (ref. 44). On April 8 and 9, 1968, the Minnesota experiment was reconfigured by a series of ground commands for this investigation. Two readings were taken in the low-energy range between 200 and 600 keV. The results proved to be consistent with an extrapolation of data measured previously in the 2- to 20-MeV range (fig. 6-25).

Anisotropies and Gradients

Although Pioneer 8's orbit takes it only from 1.0 to 1.12 AU, the Minnesota instrument is sensitive enough to estimate cosmic-ray radial gradi-

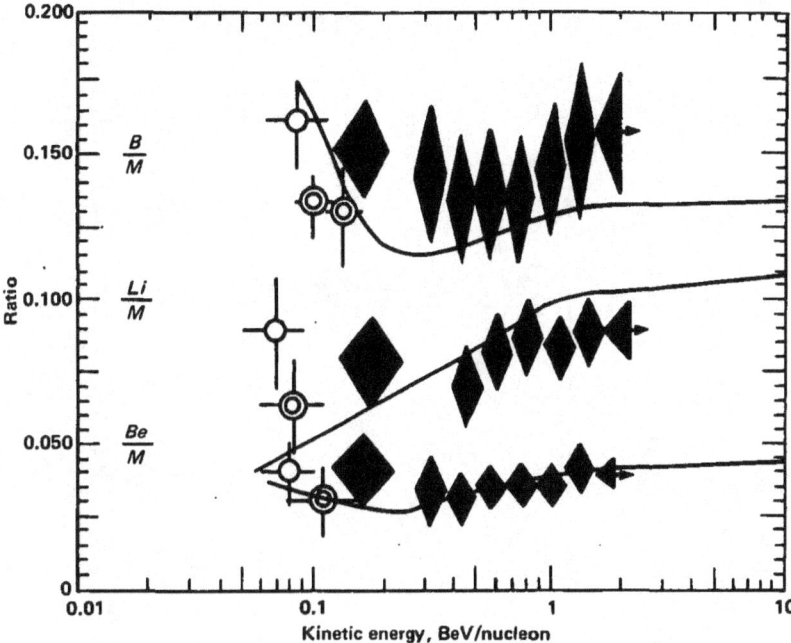

FIGURE 6-24.—The Li/M, Be/M, and B/M ratios (diamonds) as measured by Pioneer 8. The low energy measurements are from other sources. Solid lines are calculations of the energy dependences of these ratios. From: ref. 43.

TABLE 6-2.—*Chemical Abundances of Heavy Elements in the Primary Cosmic Radiation Measured by Pioneer 8*[a]

Element	Abundances Si = L	Events registered
F	0.13	(52)
Ne	1.34	(557)
Na	0.26	(110)
Mg	1.39	(575)
Al	0.30	(124)
Si	1.00	(415)
P	0.13	(53)

[a] From reference 43.

ents within the solar system. First, the instrument measured differential energy spectra of protons and helium nuclei between 40 MeV/nucleon and 2 BeV/nucleon; the analysis in this range was two-dimensional, greatly reducing the background. Second, each event was assigned to one of four quadrants, permitting a study of the anisotropies associated with the gradients. The results of these measurements are presented in

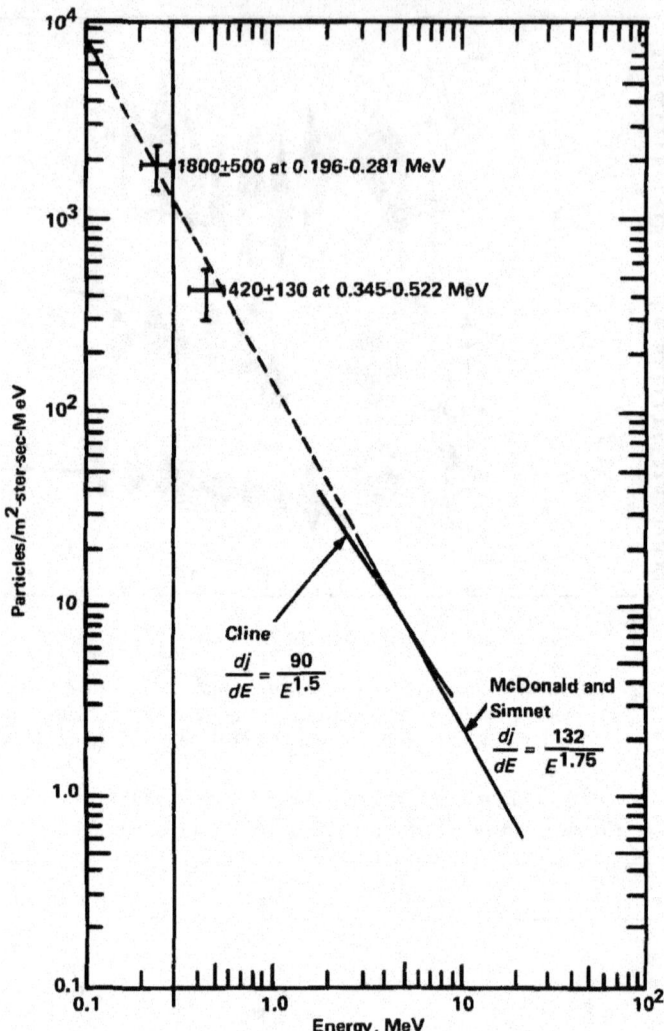

FIGURE 6–25.—Pioneer-8 measurements of low energy primary electrons. From: ref. 45.

table 6–3. In general, the cosmic ray anisotropy seems close to zero; however, it may be slightly positive in some energy ranges. The data indicate that there are no significant anisotropies above 240 MeV.

Effects of Solar Modulation

Pioneer 8 measurements of protons and helium nuclei were used in conjunction with data from balloons and terrestrial cosmic-ray monitors

TABLE 6-3.—*Gradient and Anisotropy Measurements on Pioneer 8*

Energy	Radial proton gradient	Radial proton anisotropy	Azimuthal proton anisotropy
>2 BeV	−1.5±6%	−0.31±0.28%	−0.13±0.27%
1.25 BeV to 2 BeV	0±7%	+0.26±0.45%	−0.38±0.44%
660 MeV to 1.25 BeV	+23±8%	+0.57±0.35%	−0.55±0.44%
334 MeV to 660 MeV	+28±9%	+0.36±0.38%	−0.80±0.35%
240 MeV to 334 MeV	−7±11%	+0.7±1.0%	−0.60±1.0%
63 MeV to 107 MeV	+20±15%		
>60 MeV	0±5%		
12 MeV to 25 MeV	0±25%		

to observe the solar modulation of cosmic rays during the 1965 to 1968 period (ref. 45). Pioneer 8 instrumentation covered the rigidity ranges 200 MeV to 2 BeV, for protons, and 400 MeV to 2 BeV, for helium nuclei. Figures 6-26 and 6-27 show the results from Pioneer 8 compared with similar data from other sources at earlier times. The decreases noted from 1965 to 1968 are due, of course, to rising solar activity and the solar system's increasing ability to exclude galactic cosmic rays. When these changes are compared particle-by-particle, a number of new features arise that cannot be explained by the simple diffusion-convection theory of cosmic-ray modulation. The authors note, though, that models incorporating energy-loss effects are more successful in explaining these features.

THE STANFORD RADIO PROPAGATION EXPERIMENT (ALL PIONEERS)

As described in ch. 5, Vol. II, the Stanford radio propagation experiment operates in a closed loop which employs the 150-ft paraboloidal antenna and associated transmitting equipment at Stanford University, the spacecraft receiver and transmitter, and the facilities of NASA's Deep Space Network. Basically, the experiment measures the integrated electron content between the spacecraft and the Earth. Corrections for the Earth's ionosphere are made with the help of radio-propagation measurements using Earth satellites such as the Beacon Explorers and the Applications Technology Satellites (ATSs). The integrated electron content measurements can be very revealing scientifically when solar flares occur and when the spacecraft passes near the Sun or the Moon.

The Interplanetary Electron Number Density

Based upon Pioneer-6 data taken between February 20 and April 9, 1966, the average electron number density was 8.25 cm^{-3}, with an rms

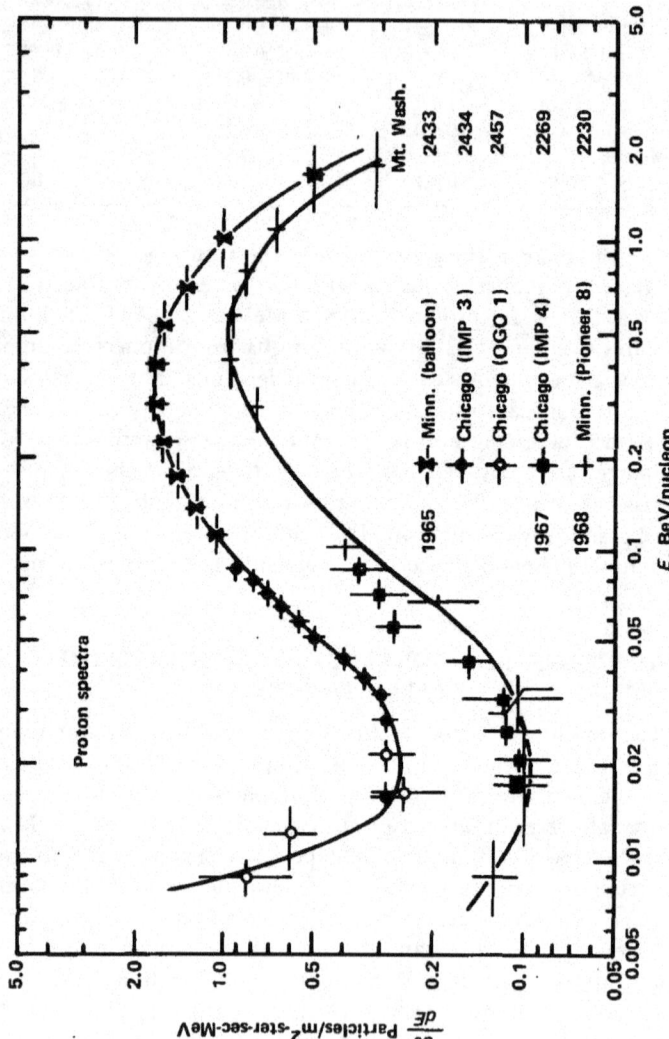

FIGURE 6-26.—Proton spectrum measured by Pioneer 8 in 1968. From: ref. 45.

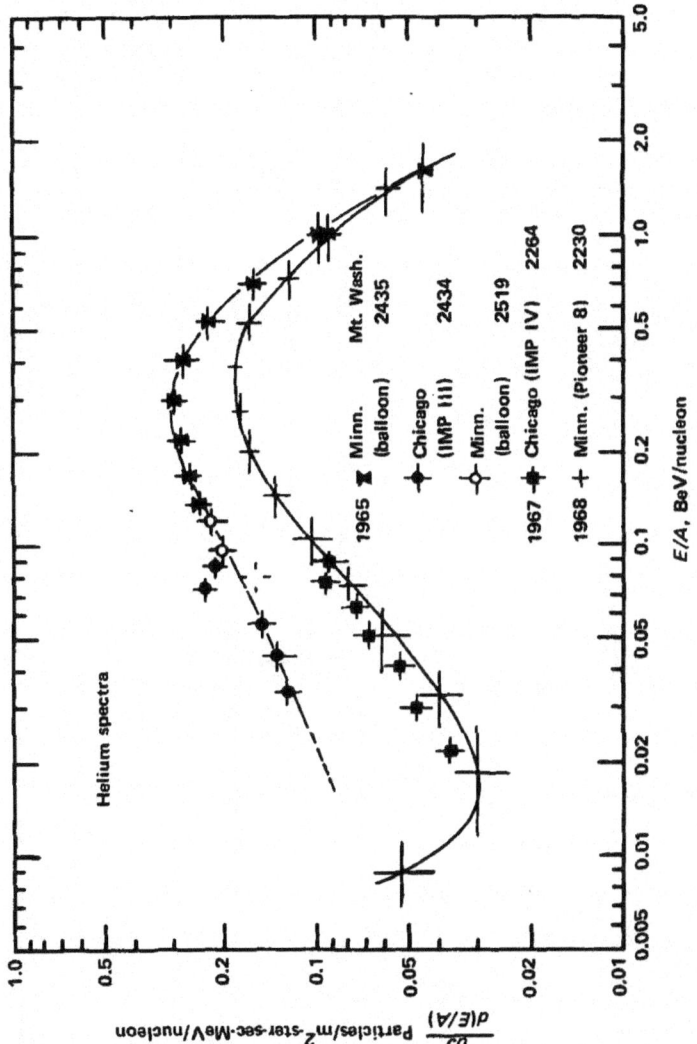

FIGURE 6-27.—Helium nuclei spectrum measured by Pioneer 8 in 1968. From: ref. 45.

of 4.43 cm^{-3} (ref. 46). As Pioneer 6 moved farther out into space, it soon became apparent that the first values reported were unusually high due to high solar activity. The spread in measured values of the total interplanetary electron content is shown for Pioneer 6 in figure 6–28. The electron number density can be computed from the slopes of the lines drawn through these scattered points. The data in the figure yield an electron number density of 5.47±4.1 cm^{-3}. A similar procedure for Pioneer-7 data leads to the value of 8.02±3.8 cm^{-3} (ref. 47).

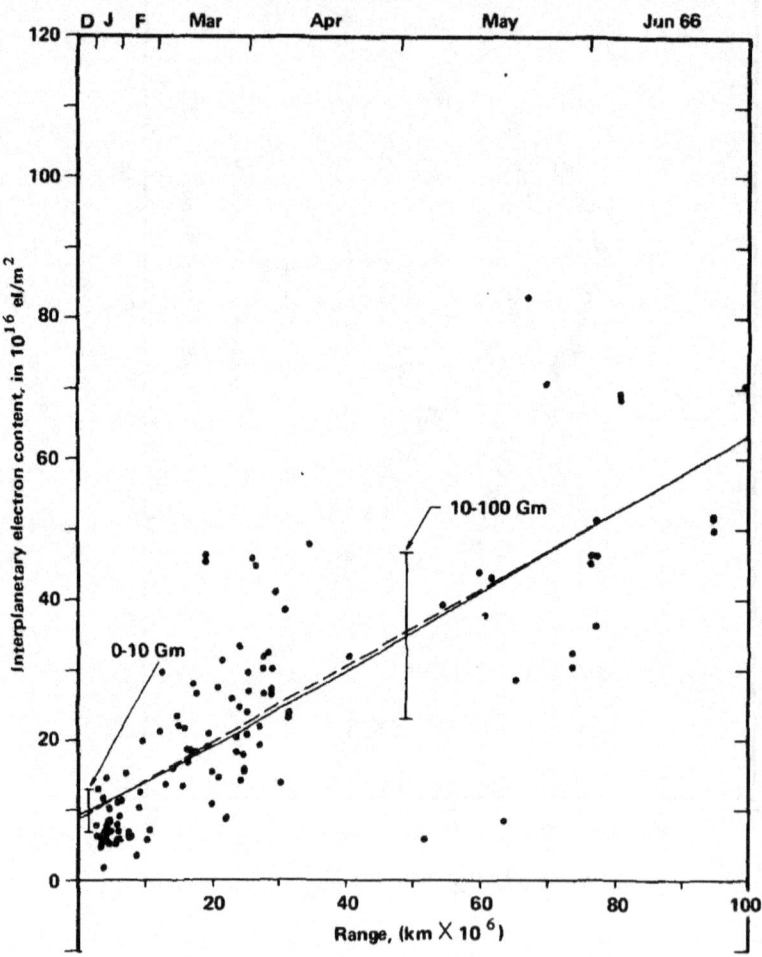

FIGURE 6–28.—Integrated electron content measured from Pioneer 6 to Earth as a function of spacecraft range. From: ref. 47.

Plasma Pulses and Clouds

The measurements plotted in figure 6-28 owe their variation primarily to changes in solar activity and, consequently, the quantity of electrons injected into interplanetary space. Some of these injections—called plasma pulses or clouds—are fairly well-defined. Many have been "mapped" after a fashion by the Stanford radio propagation experiment. Koehler has reported on the analysis of three plasma pulses occurring on October 24, 1966, November 10, 1966, and January 25, 1967 (ref. 48).

On October 24, 1966, the integrated electron content measured between the Earth and Pioneer 6 was unusually high, as shown in figure 6-29. The difference between the electron content on October 24 and October 26, a "control" day with the usual electron content, is plotted in figure 6-30. As a first approximation, the curve is triangular in shape and can be explained as due to a rectangular pulse of increased electron density travelling radially outward from the Sun and crossing the propagation path. The peak electron content was 40×10^{16} electrons/m^3. Dividing by the 10.7×10^6 km propagation path, the peak increase in electron density over the background comes to 33 electrons/cm^3. This particular pulse travelled the length of the propagation path in about 9 hr, leading to a calculated velocity of 330 km/sec. This figure corresponds well with the plasma velocity measured during the same period by the Ames plasma probe.

The event of November 10 was somewhat different in that the curve corresponding to that in figure 6-30 was flat-topped. The interpretation was that the pulse was shorter than the propagation path in this case. The flat-top, which represents a constant electron content, occurs before the leading edge reaches the spacecraft and after the trailing edge has passed the Earth.

The largest of the three pulses was noted on January 25. Its peak electron content was 56×10^{16} electrons/m^3 above the background. Indications were that this was approximately a spherical pulse 5.2×10^6 km in diameter travelling at 350 km/sec.

The Stanford group made a more detailed study of the plasma cloud ejected by the July 7, 1966, solar flare (ref. 49). Although the radio propagation experiment was being operated beyond its nominal maximum range, the description of the plasma cloud derived from the measurements is compatible with the MIT plasma probe which also measured the passage of a plasma shock at the same time (ref. 19). The shape and extent of the passing plasma cloud was calculated from the integrated electron content measured from Pioneer 6. Three cloud shapes—each deduced from a different data channel—seemed to fit the data (fig. 6-31). Each cloud model has a double structure to account for the two

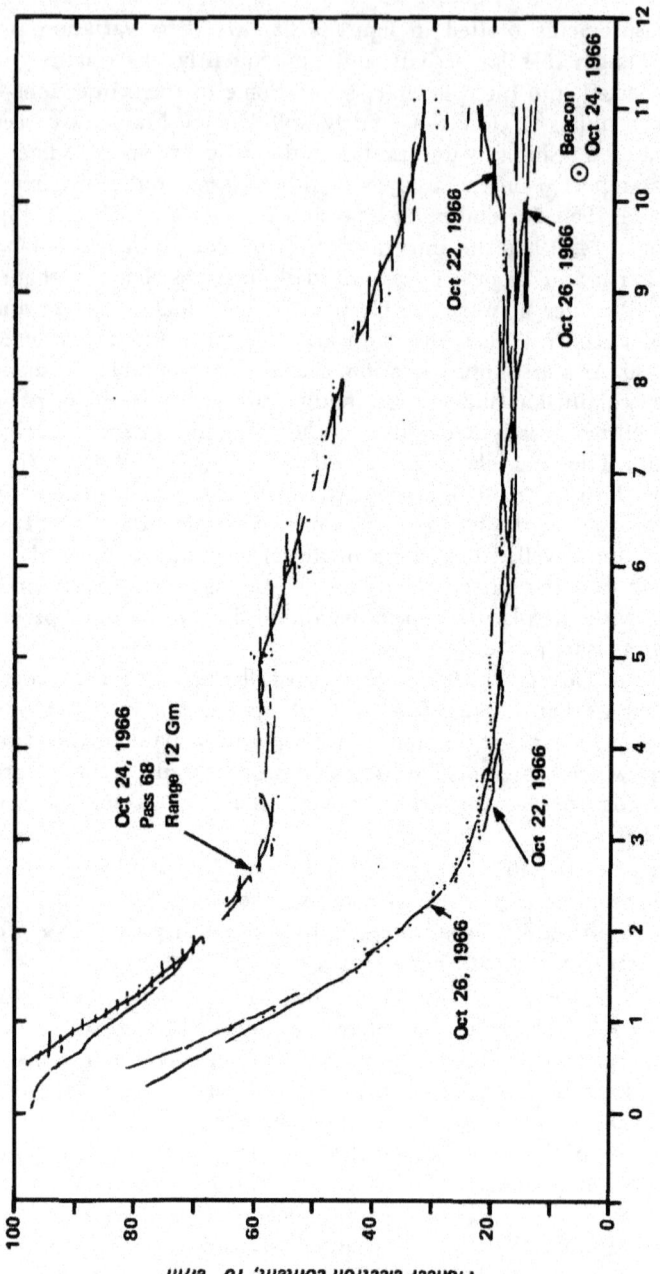

FIGURE 6-29.—The pulse of Pioneer electron content (interplanetary plus the ionospheric content) on Oct. 24, 1966. The normal content on Oct. 22 and 26 is shown for comparison. From: ref. 48.

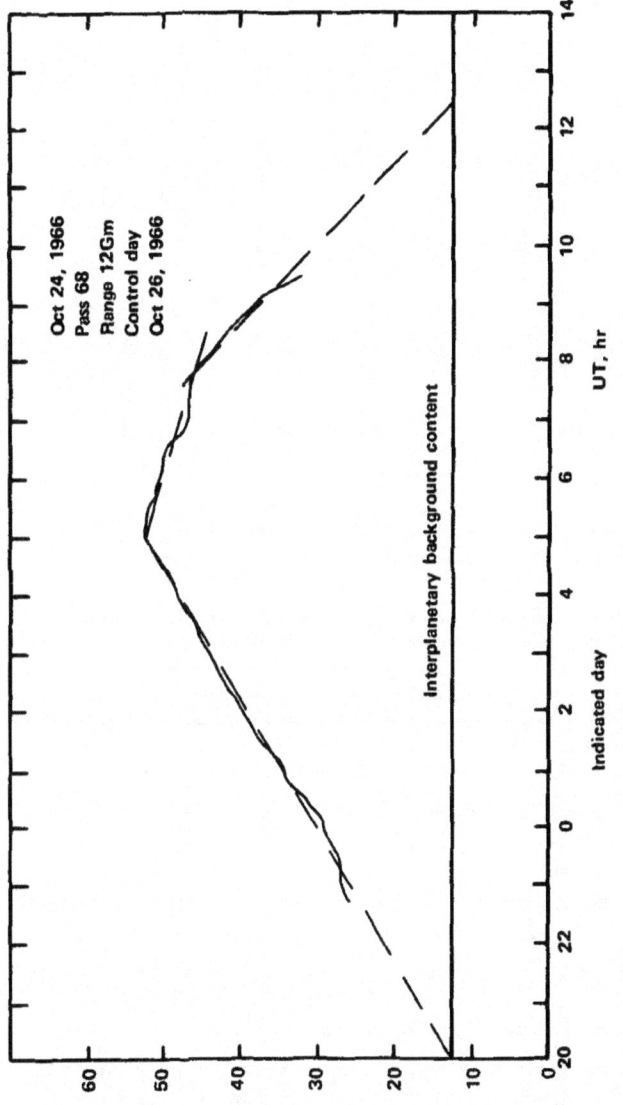

FIGURE 6-30.—The Oct. 24, 1966, pulse had a triangular shape on the electron content–time plot, inferring that it occupied the entire distance between the spacecraft and the Earth. From: ref. 48.

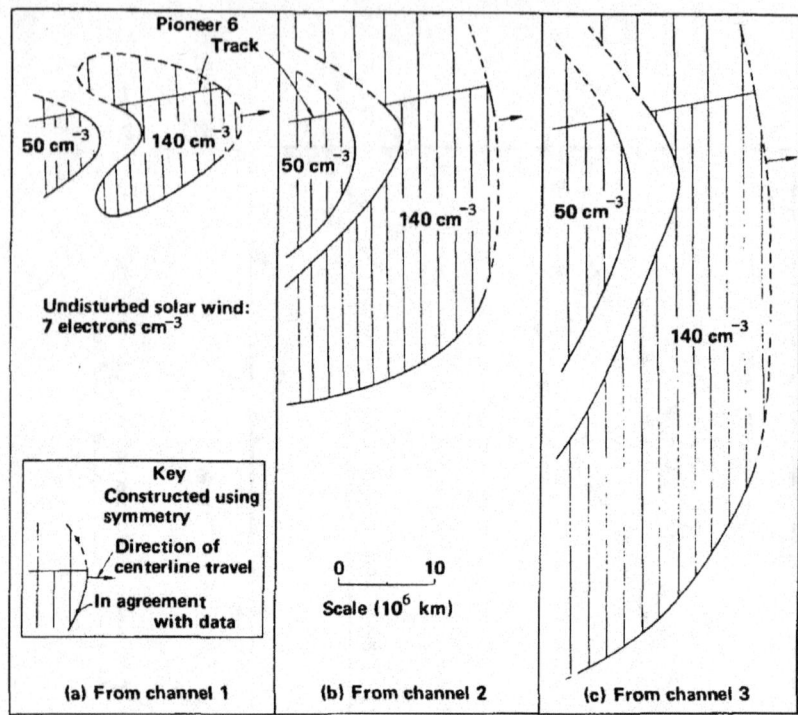

FIGURE 6-31.—Possible plasma cloud shapes. These shapes are consistent with measurements, but were restricted by simplifying assumptions and incorporate structural features based on prevailing theories about such cloud behavior. The configuration shown in (b) is considered the most likely. A gradient in density was actually measured along the Pioneer track, and a lateral gradient also probably existed; consequently, the cloud must have been broader than the outlines shown. From: ref. 49.

average density levels detected by the MIT instrument. The second cloud in figure 6-31 was thought to be the most probable configuration. Although a unique reconstruction of the cloud is impossible with the available data, the most likely models are consistent with the general conclusion that the shock fronts of the plasma clouds ejected from the Sun have radii of curvature of about 0.5 AU by the time they reach the Earth.

The January 20, 1967, Lunar Occultations

When the Moon occulted the Pioneer-7 spacecraft on January 20, 1967, radio signals sent from the 150-ft Stanford antenna were diffracted by the edge of the lunar disk and also refracted by the lunar ionosphere (ref. 50). The geometry of the situation is portrayed in figure 6-32. Of course, if there is no lunar ionosphere at all, only the classical Fresnel

SCIENTIFIC RESULTS 123

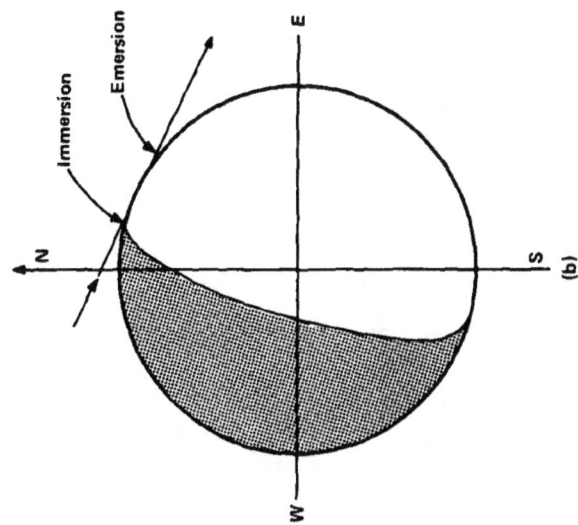

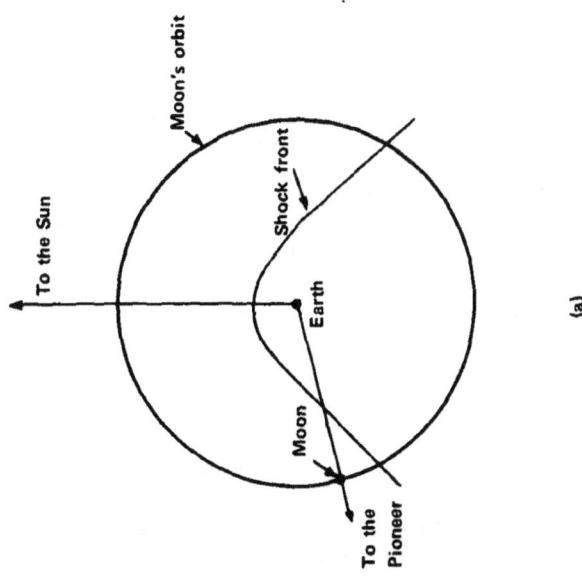

FIGURE 6-32.—(a) Relative positions of the Sun, the Moon, and Pioneer 7 in the plane of the ecliptic Jan. 20, 1967. (b) The lunar occultation of Pioneer 7 as seen from Stanford University. Immersion occurred at 05:30 UT; emersion was at 05:49 UT. From: ref. 50.

diffraction pattern will be measured (fig. 6-33). If an ionosphere is present, however, its refractive effects will displace the diffraction pattern in time. In this case, the difference in the angles of refraction for the 49.8- and 423.3-MHz signals were used to compute electron density.

The ray path from the Stanford antenna to Pioneer 7 was partially in the shadow of the Moon during immersion but was fully illuminated during emersion. The angles of refraction were -2.3 microradians and -5.7 microradians for immersion and emersion, respectively. The minus sign indicates that the electron density increases with height near the surface of the Moon, and that a tenuous ionosphere may be created —at least on the sunlit side—by the interaction of the solar wind with the lunar surface.

Solar-Wind Flow Patterns

By taking measurements during two periods each day, first using a Pioneer spacecraft ahead of Earth and then another behind it, corotating solar-wind flow patterns are clearly visible (ref. 51). The density patterns observed are not consistent with the hypothesis of steady corotating flows—there are large transients which occur too rapidly. The Stanford group has observed that identifiable features recur with nearly but not exactly the same period on successive solar rotations. Croft suggests that these patterns might be due to the corotation of thin steady streams that fluctuate in direction. These data might also indicate that some corotating regions are of low density and featureless while others are dense and highly disturbed.

RADIO PROPAGATION EXPERIMENTS USING THE SPACECRAFT CARRIER (ALL PIONEERS)

With the Stanford radio propagation experiment it is possible to measure the integrated electron density between the spacecraft and the Earth, as described in ch. 5, Vol. II. However, useful scientific information can also be obtained concerning transient space phenomena by observing changes in the Faraday rotation of the spacecraft S-band transmitter. Levy and his associates at the California Institute of Technology and the University of Southern California have used the DSN 210-ft antenna at Goldstone to measure transient Faraday rotations during solar occultation of Pioneer 6 (ref. 52). The geometry of the occultation is shown in figure 6-34. As the spacecraft line of sight approached the Sun, the S-band telemetry signal passed through increasingly dense regions of the solar corona. By November 16, however, the signal-to-noise ratio had deteriorated to the point where the experiment had to be discontinued until the signal was reacquired on November 29. At three points (marked A, B, and C on fig. 6-34) between 6 and 11 solar

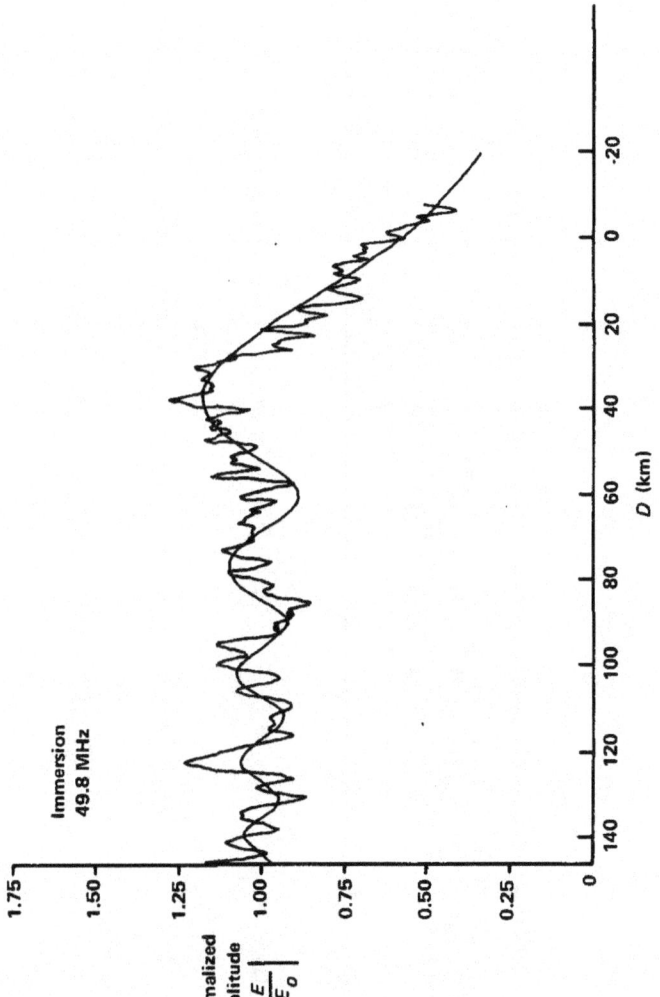

FIGURE 6-33.—Theoretical and actual diffraction pattern observed during the lunar occultation of Pioneer 7 at 49.8 MHz. From: ref. 50.

126   THE INTERPLANETARY PIONEERS

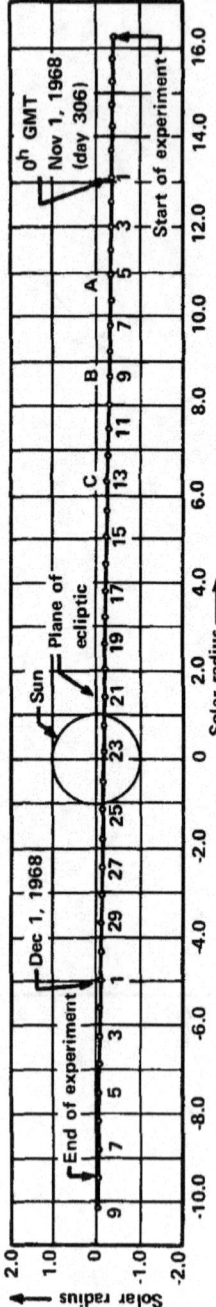

FIGURE 6-34.—Projection of Pioneer-6 orbit relative to the plane of the ecliptic. From: ref. 52.

radii, Faraday rotation transients were recorded (fig. 6-35). The duration of each event was about 2 hr. The transients were poorly correlated with solar flares, but it was noted that bursts of radio noise in the dekameter range occurred prior to the observation of the Faraday rotation phenomena.

Later studies of these transients at GSFC by Schatten led to the correlation of Levy's three events with specific Class-1 solar flares and subflares which preceded the events (ref. 53). Schatten interpreted Levy's evidence in terms of a magnetic bottle expanding from the corona to perhaps 10 to 30 solar radii before contracting back toward the corona. Resembling an elongated blister, the magnetic bottle and its electronic content would cause the observed Faraday rotation transients.

TRW SYSTEMS ELECTRIC FIELD EXPERIMENT (PIONEERS 8 AND 9)

The Pioneer Electric Field Experiment is physically associated with the Stanford Radio Propagation Experiment, utilizing its short 423-MHz antenna as a detector; but scientifically it is more closely allied with those experiments measuring characteristics of the interplanetary plasma; i.e., the Ames plasma experiment and the Ames and Goddard magnetometers. The electric field experiment was added late in the Block-II program and was thus rather limited in its allotments of weight, power, and telemetry capacity. The engineering design and physical operations of this experiment are described in ch. 5, Vol. II.

Early in the development of space physics, scientists concentrated primarily upon measuring solar plasma velocity, density, and temperature. Plasma dynamics, including the study of plasma waves and other "cooperative" phenomena, was generally not emphasized. It was recognized, of course, that waves and many other dynamic phenomena were not being detected with the usual plasma instruments. Earth satellites soon began carrying vlf (very low frequency) radio listening experiments and high-sensitivity instruments like the "LEPEDEA" analyzers. These instruments began to reveal the true complexity of the plasma environment near the Earth. It was, therefore, desirable to install a vlf or electric field experiment on the Block-II Pioneers. Fortunately, this proved possible.

Near the Earth's orbit the solar wind is very dilute, and the plasma is truly collisionless. Individual electrons and positive ions are influenced only by dc electromagnetic fields or by fields due to the organized motion of plasma particles in the form of ac plasma waves. The Pioneer Electric Field experiment was designed to detect these microscopic plasma phenomena. The overall size of the Pioneer spacecraft and its appendages is small compared to the Debye length in interplanetary

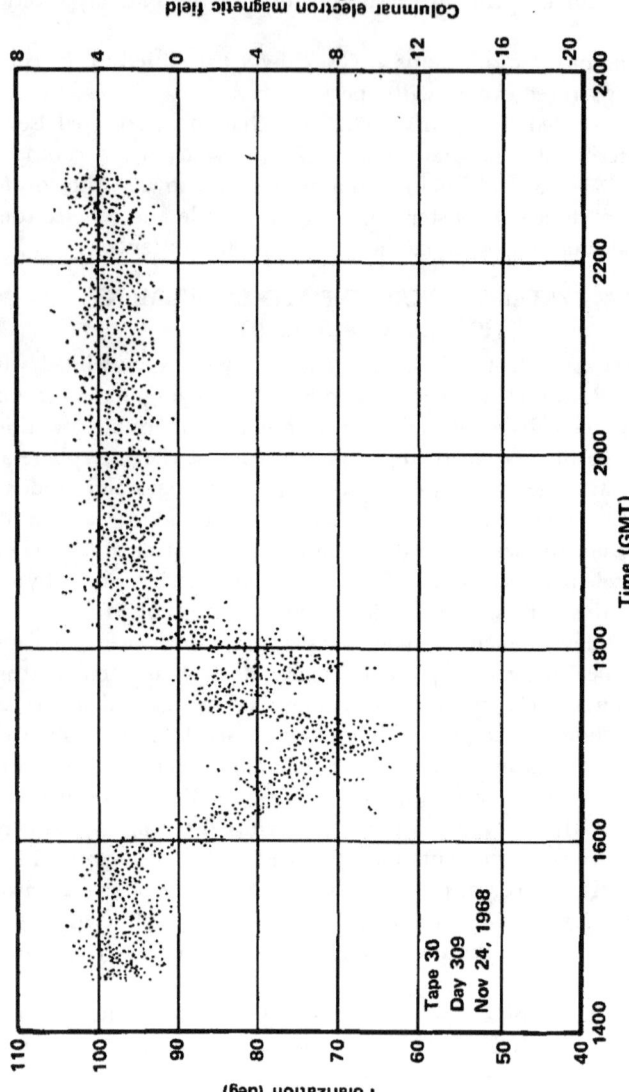

FIGURE 6-35.—One of the transient Faraday-rotation phenomena observed by Levy et al. From: ref. 52.

space and also the minimum wavelength for any undamped plasma oscillation. Thus, the spacecraft actually represents a "microscopic" measuring platform immersed in plasma phenomena of much greater fundamental size. The 423-MHz antenna is a relatively insensitive, but adequate, capacitively coupled sensor that detects plasma waves sweeping past the Pioneers in interplanetary space.

While magnetometers have helped scientists understand microscopic electromagnetic phenomena in space, the Pioneer Electric Field experiment is primarily electrostatic in nature—in fact, it was the first low-frequency (under 20 kHz) electric field experiment to be flown in deep space. The Pioneer instrument detects density fluctuations within the plasma rather than the motions of current systems indicated by magnetometers. In this sense, the electric field experiment allows us to study the plasma from an entirely different vantage point than the more conventional plasma probes and magnetometers. Electrostatic plasma phenomena can carry considerable energy in deep space and strongly affect overall plasma behavior. The following discussion of the results obtained from this experiment underscores the importance of these extremely simple, lightweight instruments in our understanding of the interplanetary medium.

Presentation of Early Results

The initial results from the Pioneer-8 Electric Field experiment were reported by Scarf et al. (ref. 54). These experimental data were treated in three categories.

Broadband measurements.—During the spacecraft's passage across the Earth's magnetosphere, very low amplitude vlf oscillations were detected. On December 14, Pioneer 8 first encountered the streaming plasma in the distant magnetosheath, as indicated in figure 6–36. The electric field experiment detected some plasma waves before the Ames plasma probe registered its first bursts of plasma around 2140 UT. Apparently the crossing of the magnetosphere was completed about 0230 on December 15 when both the electric field experiment and the Ames plasma probe indicated enhanced activity. Almost coincident with the penetration of the magnetosheath, Earth-based magnetic field instrumentation reported a magnetic disturbance (a sudden commencement). Within minutes of the terrestrial indication, the Pioneer electric field experiment also detected the disturbance. Evidently, the phenomena stimulating terrestrial magnetic storms also intensify interplanetary plasma waves.

The 400-Hz channel.—When the preceding broadband data are combined with information from the narrow-bandwidth 400-Hz channel,

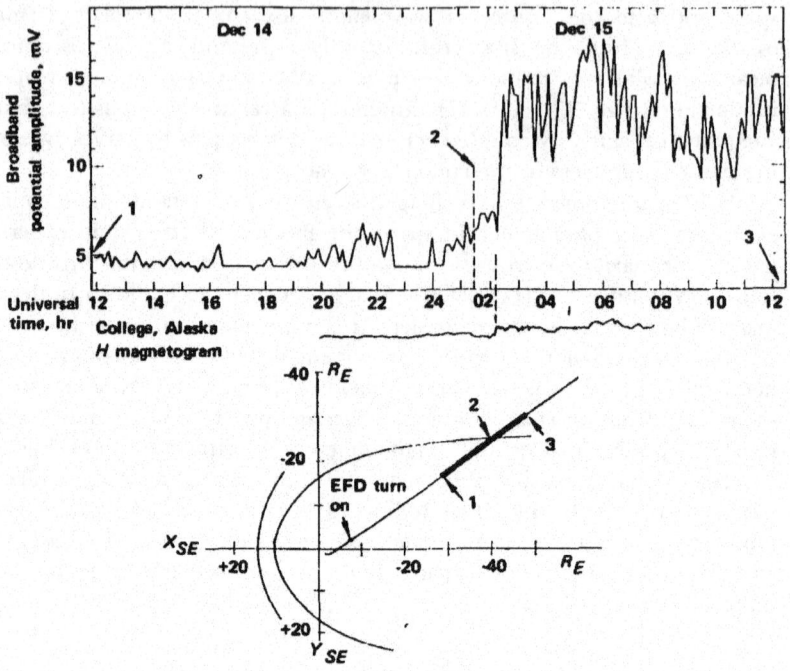

FIGURE 6-36.—Broadband wave amplitudes in the outer magnetosphere and magnetosheath. The projection of the near-Earth trajectory in the ecliptic plane is shown, and the heavy segment represents the period from 1200 UT, Dec. 14 to 1210 UT on Dec. 15 (indicated by the numbers 1 and 3). Point 2 shows where the Ames Research Center plasma probe first started to detect continuous streaming plasma (personal communication), but bursts were encountered earlier. The H component of the College, Alaska, magnetogram shows a sudden commencement at 0215 UT, Dec. 15, followed by a storm (this ssc was categorized as a sudden impulse by some observatories), and the broadband wave amplitude rose shortly thereafter. From: ref. 54.

one obtains a measure of the spectral width of the low-frequency noise band.

Telemetry from the 400-Hz channel revealed a regular modulation that was quickly associated with the spacecraft's spin rate, more precisely with the instantaneous position in space of the Stanford antenna. The effect was greatest when this antenna was pointed toward the Sun. Apparently, the physical cause of the modulation is a Sun-aligned space-charge cloud surrounding the non-conducting spacecraft.

Despite the modulation, the 400-Hz channel is clearly sensitive to the plasma waves detected by the broadband channel. Further conclusions were not drawn at the time this initial paper was written.

The 22-kHz channel.—The narrow-bandwidth 22-kHz channel provides information about plasma oscillations when the electron concentration is relatively low. There is a natural noise background at this frequency, but it usually lies well below the experiment's threshold. Rarely, however, intense bursts of 22-kHz noise activate the receiver. The noise burst portrayed in figure 6–37 is typical. Although rare, these noise bursts do not appear to be random, being weakly correlated with solar and geomagnetic activity and strongly correlated with proximity to the Earth's magnetosphere.

The following tentative observations were presented by Scarf et al. (ref. 54):

(1) Even when the Sun is quiet, low-frequency electric waves ($>100Hz$) can be detected in the solar wind although the lowest levels are near the in-flight background.

(2) Wave amplitudes at the lowest frequencies vary markedly with changing conditions in interplanetary space. These electric field changes are correlated with local changes in the plasma environment as registered on the Ames plasma probe.

(3) As Pioneer 8 moved away from the Earth, the effects of corotation and solar-wind travel times were evident when comparing disturbances recorded at both the Earth and the spacecraft.

(4) Large-amplitude high-frequency waves, detected when the spacecraft was far from Earth, are apparently the result of bursts of interplanetary, but Earth-associated, electron oscillations.

Shock Structures

At the 1969 Summer Advanced Study Institute at the University of California at Santa Barbara, further results were presented on the shock structures detected by the Pioneer electric field experiment (ref. 55). Data from Pioneers 8 and 9 and OGO 5 were used to demonstrate the several types of shock structures found in the high Mach-number solar plasma colliding with the Earth's magnetosphere. The most common structure reported was a large-amplitude magnetohydrodynamic pulse having a characteristic length equal to the initial gradient and a trailing wavetrain. Energy in these shock structures is apparently dissipated via electrostatic wave turbulence which arises from instabilities. Further thoughts concerning these interactions were presented in a second paper at this same meeting by Scarf and his associates (ref. 56).

Measurements in the Distant Geomagnetic Tail

The plasma-probe and electric-field data recorded as Pioneer 8 crossed the Earth's geomagnetic tail during January 1968 were reported in 1970

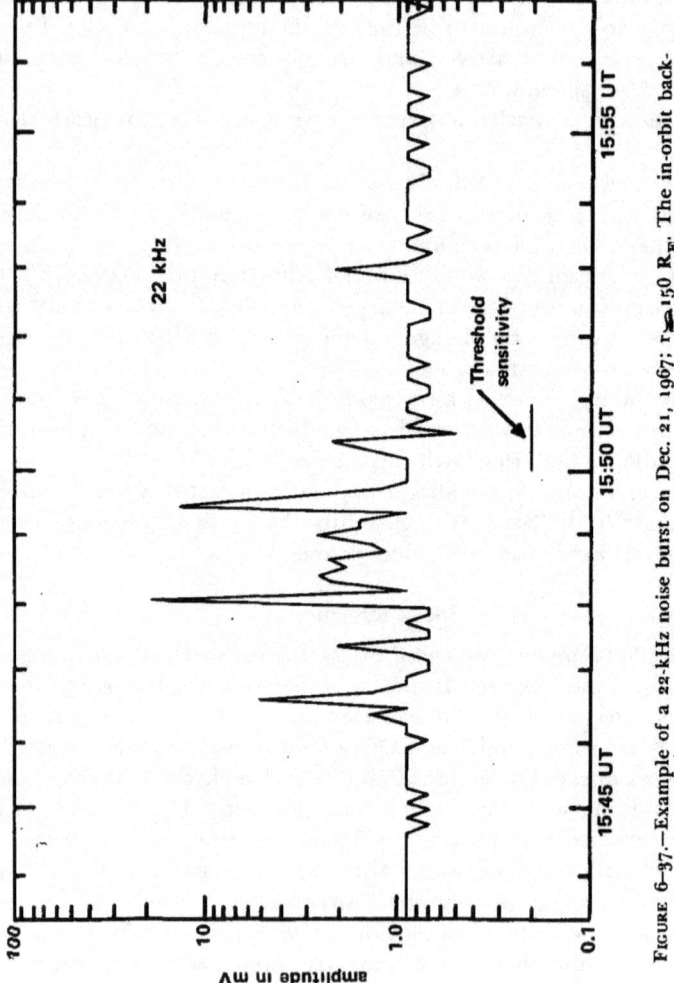

FIGURE 6-37.—Example of a 22-kHz noise burst on Dec. 21, 1967; $r \approx 150$ R_E. The in-orbit background is near 0.68–0.89 mV, and deviations below this are found only when the most intense enhancements are also detected. These large-amplitude 22-kHz bursts are observed very rarely. From: ref. 54.

by the researchers at Ames and TRW Systems (ref. 57). Both instruments recorded disturbances near the tail boundaries between 500 and 800 Earth radii downstream. The major conclusion of this paper was that tail breakup and field-line-reconnection phenomena begin within 500 Earth radii.

Multi-Instrument Correlation of Space Disturbances

The initial results from the Pioneer-8 electric field experiment showed clearly their close correlations with terrestrially detected magnetic activity. Because the other Pioneer instruments also record space events (from a different perspective), correlations between different onboard instruments should also be obvious in many instances. Scarf, in his 1970 review paper, illustrated a three-way correlation during a Forbush decrease. Figure 6–38 indicates how the Pioneer-8 magnetometer, electric-field experiment, and the Minnesota cosmic-ray experiments all recorded the same event.

Another interesting correlation between different instruments (on different spacecraft this time) was revealed by Siscoe et al. (ref. 58). They illustrated the striking correlation between Pioneer-8 electric-field data and the solar-wind parameters recorded by Explorer 35 (in lunar orbit) in late February 1968 (fig. 6–39). Siscoe et al. noted that the electric-field noise data are of two types: (1) bursts or spikes lasting less than 10 sec, and (2) persistent signals typically lasting a day or more. The first type of data coincide with plasma and magnetic-field discontinuities, whereas the latter are available for comparison. The persistent signals, on the other hand, correlate loosely with solar-wind density, whether the density increases are due to interplanetary shocks (the so-called "snow plow" effect) or other processes.

This multi-spacecraft correlation study also proved of value in defining the spatial extent of the Earth's influence. Siscoe and his associates showed that a huge wake region surrounds the distant geomagnetic tail (ref. 59). This analysis indicated that Pioneer 8 did not encounter undisturbed solar wind for several months following launch. In a later paper, Siscoe used this fact to explain the early anomalous E-field observations (ref. 59).

In later papers, new types of correlation studies were presented (refs. 60 and 61). Nearly simultaneous wave observation from OGO 5 and Pioneer 9 were compared and used to provide an in-flight calibration for the simple Pioneer instrument. Analysis of these wave observations suggested that the wave spectrum varies with radial distance from the Sun.

In concluding the discussion of the electric-field experiment, it should be noted again that the experiment had only very limited telemetry

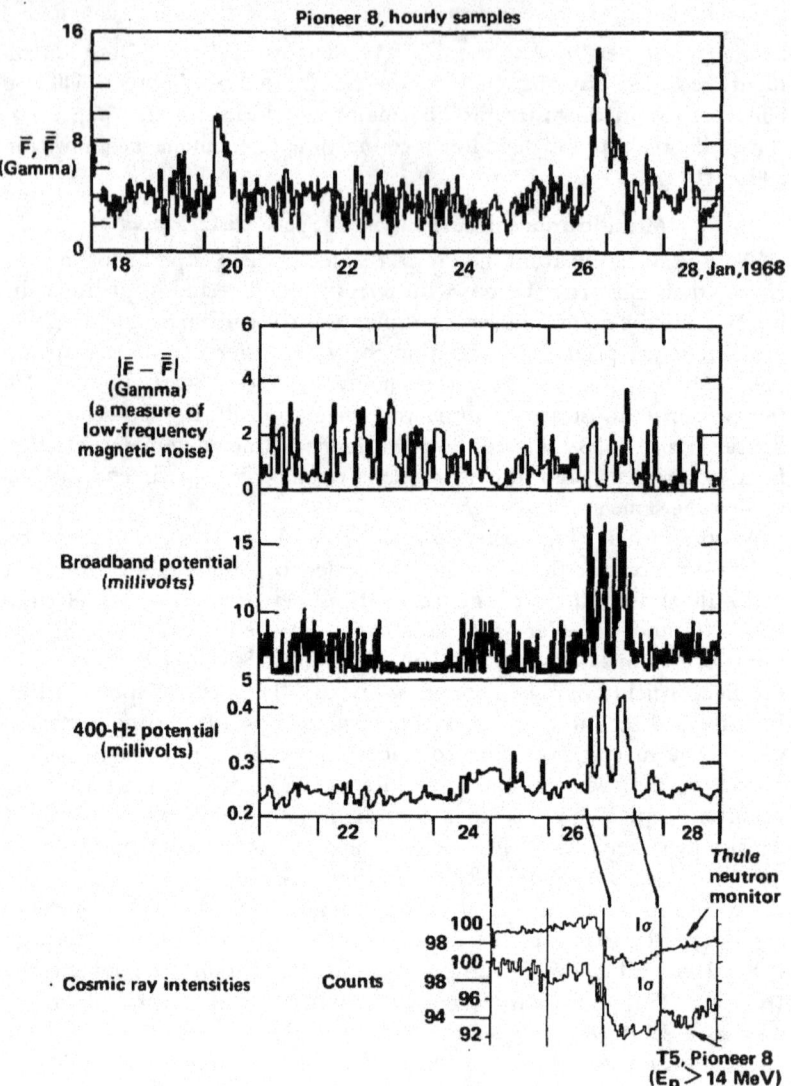

FIGURE 6-38.—Pioneer 8 magnetometer data (top) and electric-field data (middle) reveal interplanetary shock. Cosmic-ray readings (bottom) show attendant Forbush decrease.

capacity assigned, and that the interpretation of results is somewhat complicated by interactions of the interplanetary medium with the spacecraft. For example, the experiment was too limited for the unambiguous determination of plasma wave modes in interplanetary space.

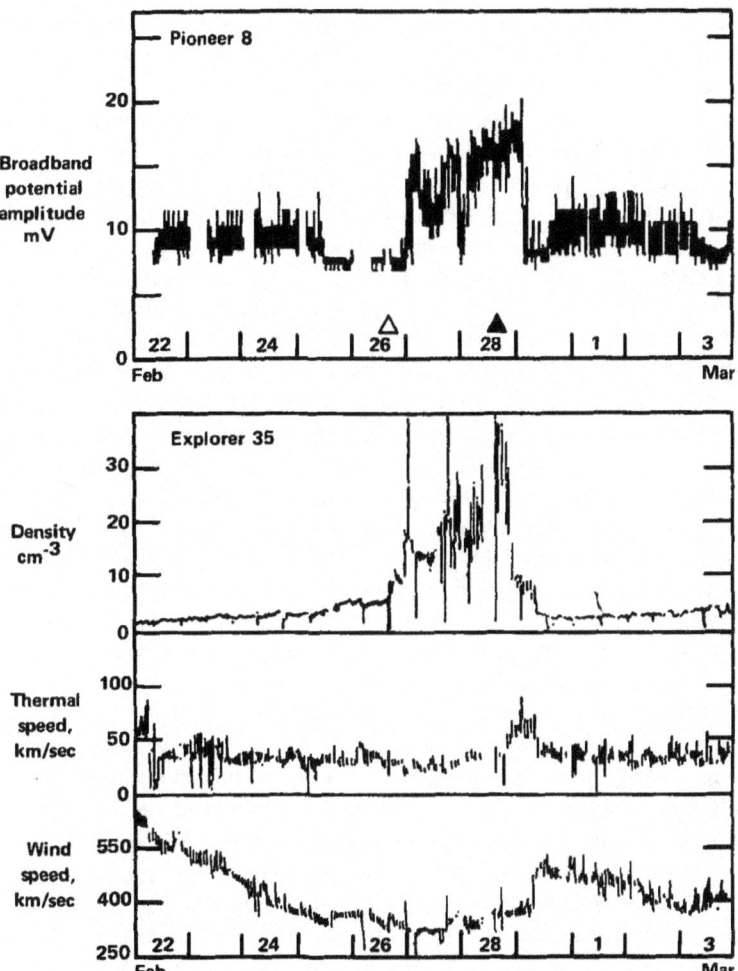

FIGURE 6-39.—Interplanetary shock registered almost simultaneously on Pioneer 8 electric field experiment and Explorer 35 plasma probe.

THE GODDARD COSMIC DUST MEASUREMENTS (PIONEERS 8 AND 9)

During the early days of the Space Age, cosmic dust was thought to be a serious hazard to men and machines operating outside the Earth's protective atmosphere. More accurate measurements of cosmic dust particles have since shown these fears to have been unwarranted. Sensitive external surfaces on long-lived Earth satellites may suffer some degradations, but neither manned nor unmanned spacecraft have been compromised. Nevertheless, cosmic dust particles do exist, and their presence in space demands a scientific explanation.

Are cosmic dust particles products of cometary disintegration or the debris from collisions within the asteroid belt? Most of our insight into this question at present comes from ground-based photographic and radar measurements of meteor trails. These data suggest that almost all cosmic dust trajectories are heliocentric with the orbital characteristics of comets rather than asteroids. Further, the particles seem "fluffy" and of low density. The Pioneer cosmic dust experiment, which flew on Pioneers 8 and 9, was designed to help answer this question of particle origin with *in situ* data from deep space.

The experiment itself (described in detail in Vol. II) was designed and tested by a team of scientists and engineers at Goddard Space Flight Center. Arrays of plasma and acoustical sensors can measure particle directions, energies, and, through a time-of-flight technique, velocities. The early results from Pioneer 8 described below have been reported by the Goddard group (ref. 62).

During the first 390 days of continuous exposure of the Pioneer-8 sensors, numerous events (several per day) were recorded by the front sensor array alone, the rear sensor array alone, or the microphone sensor alone. These data were not completely analyzed at the time the referenced papers were written. However, six time-of-flight events involving both front and rear sensor arrays were also registered. These are considered highly important to the question of cosmic dust origin because orbital information can be derived from the measurements.

The six time-of-flight events in a space of 390 days represent a rate 3.8×10^4 lower than the rate recorded by a time-of-flight experiment on OGO 1. It is surmised that the high OGO-1 rate was due to coincident noise pulses in that experiment—noise was a serious problem with early scientific satellite cosmic dust experiments. Early Pioneer-8 results confirm expectations from zodiacal light measurements.

From a knowledge of the spacecraft trajectory and orientation at the instant of each event and the telemetered data indicating times of flight and the specific sensors activated in the front and back arrays, it was possible to derive the particle orbits shown in table 6–4. These data indicate a cometary origin for the six particle events, reinforcing the conclusions derived from ground-based observations.

The most interesting of the six events reported in the two papers occurred on April 13, 1968. Apparently, one front sensor segment and two rear sensors responded, inferring that the particle partially disintegrated upon first impact, showering the rear array with a conical spray

TABLE 6-4.—*Particulars for the First Six Time-of-Flight Events Detected by Pioneer 8*[a]

Date	MPS	SCS	MPI	TPS	PM	a	e	q	q'	ω	Ω	i	π
3/11/68	7.5	27.3	0±27	19.3	2.1	0.67	0.58	0.28	1.07	—	—	0	339
3/26/68	18.0	28.8	0±27	12.5	0.93	0.58	0.84	0.095	1.07	—	—	0	7
4/13/68	27.6	28.5	0±27	16.05	2.2	0.62	0.97	0.019	1.22	—	—	0	24
4/14/68	34.0	28.4	0±27	9.06	0.78	0.55	0.97	0.013	1.09	—	—	180	17
5/25/68	15.7	27.9	0±27	13.28	1.2	0.59	0.81	0.11	1.07	—	—	0	50
1/24/69	5.9	30.4	33^{+8}_{-4}	31.5	1.9	1.28	0.24	0.98	1.59	49.9	102	5.9	152

MPS—Measured Particle Speed (km/sec)
SCS—Spacecraft Speed (km/sec)
MPI—Measured Particle Inclination (deg)
TPS—True Particle Speed (km/sec)
PM—Particle Mass (derived) (gm × 10^{-11})
a—Semi-Major Axis (in AU)
e—Eccentricty

q—Perihelion Distance (in AU)
q'—Aphelion Distance (in AU)
ω—Argument of Perihelion (deg)
Ω—Longitude of the Ascending Node, deg
i—Inclination of the Orbital Plane to the Ecliptic, deg
π—Longitude of Perihelion (deg)

[a] See reference 63.

of debris. No such fragmentation was observed during laboratory tests with particles fired from an electrostatic accelerator. In view of the possible friable nature of cosmic dust material, this type of event was not unexpected. The much higher rates of solitary front and back array events also tend to indicate fluffy particles with poor penetrating powers.

The April 13, 1968, event was notable in two other aspects: (1) its impact energy exceeded 80 ergs, more than any other particle recorded; and (2) it was the only particle that activated the acoustical sensor. Thus, independent measurements of the particle mass were possible from the energy and momentum equations. These were 2.3×10^{-11} and 1.6×10^{-11} grams—relatively good agreement for this kind of experiment. From this information an orbit for the particle can be computed (fig. 6-40). However, it is cautioned that the response of the acoustic sensor to the postulated spray of debris is unknown and was assumed to be the same as that of a solid particle.

Summarizing, the early Pioneer-8 cosmic dust data tend to confirm the hypotheses that cosmic dust is almost exclusively of cometary origin and rather fluffy or friable in character.

In a paper presented at the XIIIth Plenary Meeting of COSPAR in May 1970, Berg and Gerloff summarized Pioneer results to that date (ref. 63).

(1) The microphone micrometeoroid detectors employed on many early spacecraft also respond to the cosmic rays generated by solar flares. This effect was responsible for the much higher micrometeoroid fluxes "measured" by these craft in the early days of space science.

(2) The micrometeoroid flux between 0.7 and 1.1 AU is $2\pm0.5 \times 10^{-4}$ particles/m^2-sec-2πsr and shows a cutoff at a mass of 5×10^{-12} g (fig. 6-41).

(3) Among the eight detected particles for which orbits could be computed were two that travel in the orbital planes of known comets (Encke and Grigg-Skellerup).

THE PIONEER CELESTIAL MECHANICS EXPERIMENT (ALL PIONEERS)

All spacecraft launched out of the Earth's gravitational "well" provide opportunities for improving solar system constants and ephemerides. Although the Pioneer spacecraft did not pass close to any other planets, their trajectories were affected by the Moon. Further, the launch of four similar spacecraft, of known mass and equipped with tracking aids, into heliocentric orbits made possible more accurate determinations of the Astronomical Unit (AU) as well as the Earth's ephemeris.

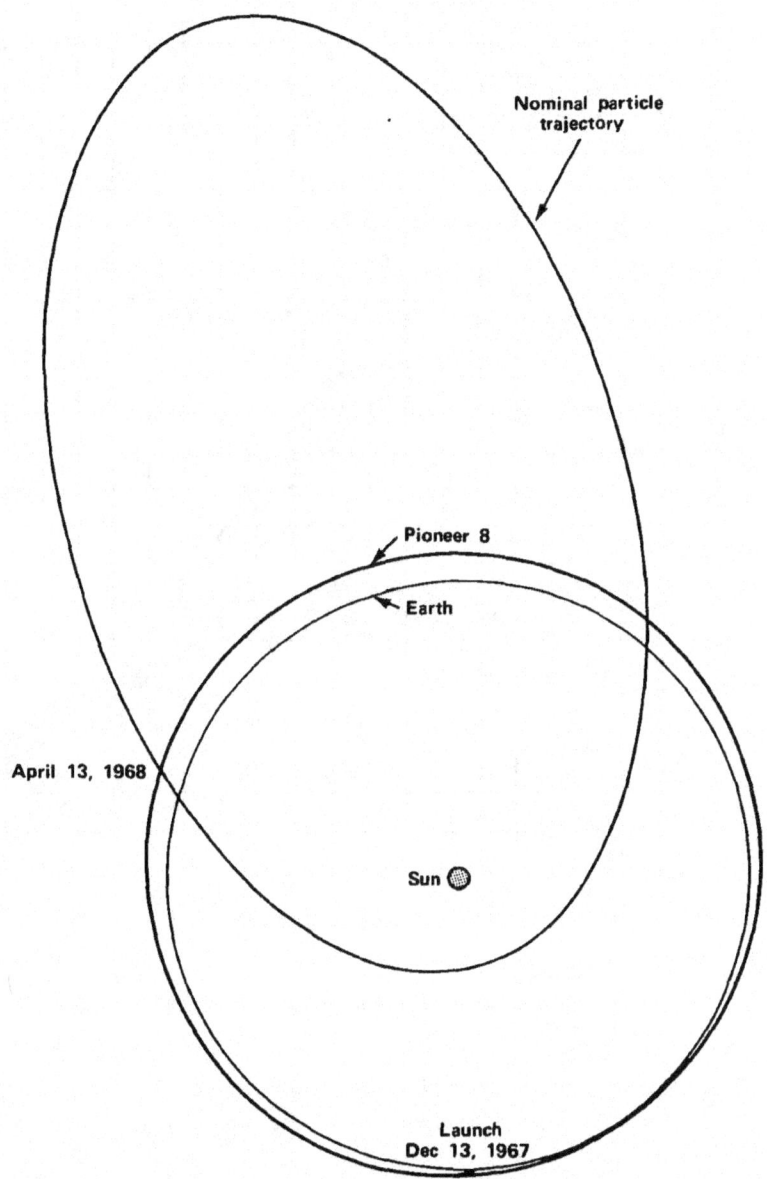

FIGURE 6-40.—Postulated orbit for the particle recorded on Apr. 13, 1968. This was a time-of-flight event. From: Berg, 1969.

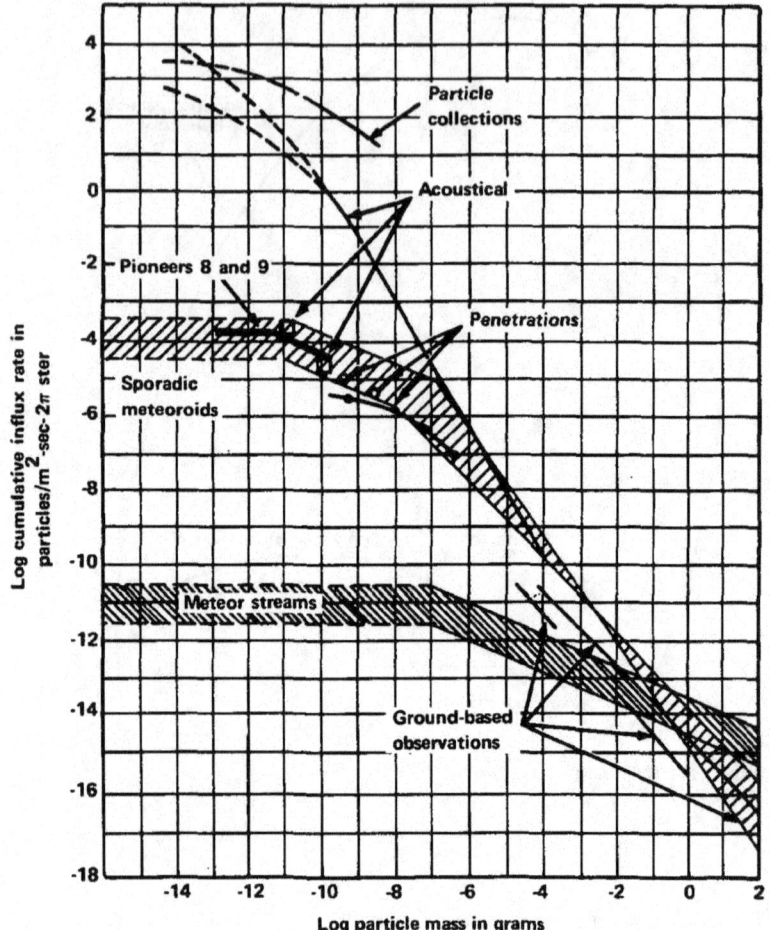

FIGURE 6-41.—Micrometeoroid fluxes as functions of mass. The heavy lines represent experimental data, while the shaded areas are theoretical. Pioneer data agree well with theory. From: ref. 63.

The three formal objectives of the experiment were:

(1) Obtain primary determinations of the masses of the Moon and Earth and of the AU

(2) Improve the ephemeris of the Earth

(3) Investigate the possibility of a General Relativity test, using Pioneer orbits and data (ref. 64).

The source of all data for this experiment is the Deep Space Network. The spacecraft carried no special equipment for this experiment. DSN two-way coherent Doppler data, judged as "good" by the tracking stations, were put on magnetic tapes. The tapes also contained mission data, such as the specific DSN station responsible for each group of data. Obvious mistakes were edited from the tapes.

Although the Pioneers, unlike the Mariners, made no midcourse maneuvers, they did carry out the two types of orientation maneuvers described in Vol. II. Tracking data indicate that some trajectory perturbations resulted from these maneuvers. Fortunately, the orientation maneuvers were usually concluded within a day or two of launch. From the tracking data on Pioneer 7, it is apparent that the Type-II orientation maneuver executed on August 19, 1966, changed the spacecraft velocity only by a few hundred millimeters per second. More serious was the gas leak on Pioneer 6, which provided an unknown source of uncompensated momentum transfer—in other words, a source of thrust that confused the analyses. The gas leak rendered the first 6 months of Pioneer-6 tracking data useless in terms of the objectives of the experiment.

In the first paper published on the Pioneer celestial mechanics experiment, Anderson and Hilt (ref. 64) reported the following preliminary Earth-Moon data:

$$\text{Geocentric gravitational constant} = GE = 398601.5 \pm 0.4 \text{ km}^3/\text{sec}^2$$
$$\text{Lunar gravitational constant} = GM = 4902.75 \pm 0.12 \text{ km}^3/\text{sec}^2$$
$$\text{Earth-Moon mass ratio} = \mu^{-1} = 81.3016 \pm 0.0020$$

The uncertainties were believed to represent realistic standard errors. The authors reported that most of the systematic errors in the determination of μ^{-1} probably came from the single-precision numerical computations performed on an IBM 7094. These computations were being converted to double-precision as the paper was being written.

SOLAR WEATHER MONITORING

Solar events strongly affect all that transpires in interplanetary space from the edges of the Sun's extensive corona to well beyond the Earth's orbit. The chief carriers of solar disturbances are the solar plasma (solar wind), the Sun's magnetic field, and the bursts of cosmic rays that often accompany solar activity. The interactions of these phenomena with the Earth are not obvious as far as the ordinary citizen is concerned. Rarely, he will read or hear that intense magnetic storms triggered by solar activity are hampering long distance communication, but in general the Earth is well insulated from any obvious effects of solar activity by its magnetosphere and atmosphere.

During intense magnetic storms, while most people go about their business unknowing and unconcerned, solar-induced electromagnetic effects wreak havoc with radio and long-landline communication. High frequency radio links depending upon forward- or back-scattering processes in the upper atmosphere are often impossible to use. There is also evidence that magnetic storms sometimes open circuit breakers and cause other disruptions in large electric power grids, particularly in the northern latitudes. Companies engaged in searching for minerals with magnetic detectors are often forced to suspend operations. Accurate forecasts of magnetic storms are useful (and worth money) in the sense that preparations can be made for use of other communication circuits and for otherwise reducing the impact of the storm.

Because of these terrestrial effects, several groups are interested in "solar weather"; i.e., the status of the interplanetary magnetic field, plasma fluxes, and cosmic radiation levels. The interest transcends pure science. NASA, for example, is concerned with solar events that might compromise manned space missions, particularly those that leave the shelter of the Earth's magnetosphere. The Environmental Science Services Administration (ESSA) desires advance information on magnetic storms and the injection of new charged particles into the Earth's belt of trapped radiation. These are the events that sometimes upset terrestrial communications and have some not-so-well-understood effects on the planet's weather. The Department of Defense (DOD) has similar interests for military reasons.

Pioneer solar weather reports began in January 1967. Usually, they are sent once a day to ESSA's Space Disturbance Forecast Center at Boulder, Colorado, to DOD's NORAD, and to other agencies. However, when manned flights are imminent, reports are sent hourly to NASA's Apollo Mission Control Center at Houston, Texas. The reports include:

(1) The corotation delay, the expected time in days between the measurement of a disturbance at the spacecraft and its arrival at Earth
(2) Solar-wind velocity, density, and temperature
(3) Cosmic-ray intensities in several energy bands
(4) The general condition of the interplanetary magnetic field.

REFERENCES

1. BURLAGA, L. F.; AND NESS, N. F.: Macro and Microstructure of the Interplanetary Magnetic Field. Can. J. Phys., vol. 46, 1968, p. S962.
2. NESS, N. F.; SCEARCE, C. S.; AND CANTARANO, S. C.: Preliminary Results from the Pioneer 6 Magnetic Field Experiment. J. Geophys. Res., vol. 71, Jul. 1, 1966.

3. NESS, N. F.: Simultaneous Measurements of the Interplanetary Magnetic Field. J. Geophys. Res., vol. 71, Jul. 1, 1966, p. 3319.
4. BURLAGA, L. F.; AND NESS, N. F.: Tangential Discontinuities in the Solar Wind. Solar Phys., vol. 9, Oct. 1969, p. 467.
5. NESS, N. F.; SCEARCE, C. S.; AND CANTARANO, S. C.: Probable Observations of the Geomagnetic Tail at 10^3 Earth Radii by Pioneer 7. J. Geophys. Res., vol. 72, Aug. 1, 1967.
6. FAIRFIELD, D. H.: Simultaneous Measurements on Three Satellites and Observation of the Geomagnetic Tail at $1000R_E$. J. Geophys. Res., vol. 73, Oct. 1, 1968, p. 6179.
7. MARIANI, F.; AND NESS, N. F.: Observations of the Geomagnetic Tail at 500 Earth Radii by Pioneer 8. J. Geophys. Res., vol. 74, Nov. 1, 1969, p. 5633.
8. BURLAGA, L. F.: Directional Discontinuities in the Interplanetary Magnetic Field. Solar Phys., vol. 1, Apr. 1969, p. 54.
9. BURLAGA, L. F.: Micro-scale Structures in the Interplanetary Medium. Solar Phys., vol. 4, May 1968, p. 67.
10. SARI, J. W.; AND NESS, N. F.: Power Spectra of the Interplanetary Magnetic Field. NASA TM-X-63428, 1968. (Also available in Solar Phys., vol. 8, Jul. 1969.)
11. MARIANI, F.; BAVASSANO, B.; AND NESS, N. F.: Magnetic Field Fluctuations on the Magnetosheath Observed by Pioneers 7 and 8. NASA TM-X-63762, 1969.
12. NESS, N. F.; AND TAYLOR, H. E.: Observations of the Interplanetary Magnetic Field, July 4-12, 1966. NASA TM-X-55842, 1967.
13. MARIANI, F.; BAVASSANO, B.; AND NESS, N. F.: Interplanetary Magnetic Field Measured by Pioneer 8 during the 25 February 1969 Event. Intercorrelated Satellite Observations Related to Solar Events. V. Manno and D. E. Page, eds., D. Reidel Publishing Co., Dordrecht, 1970, pp. 427-435.
14. LAZARUS, A. J.; BRIDGE, H. S.; AND DAVIS, J.: Preliminary Results from the Pioneer VI M.I.T. Plasma Experiment. J. Geophys. Res., vol. 71, Aug. 1, 1966, p. 3787.
15. SISCOE, G. L.; TURNER, J. M.; AND LAZARUS, A. J.: Simultaneous Plasma and Magnetic-Field Measurements of Probable Tangential Discontinuities in the Solar Wind. Solar Phys., vol. 6, Mar. 1969, p. 456.
16. SISCOE, G. L.; GOLDSTEIN, B.; AND LAZARUS, A. J.: An East-West Asymmetry in the Solar Wind Velocity. J. Geophys. Res., vol. 74, Apr. 1, 1969.
17. SISCOE, G. L.; FORMISANO, V.; AND LAZARUS, A. J.: Relation Between Geomagnetic Sudden Impulses and Solar Wind Pressure Changes—An Experimental Investigation. J. Geophys. Res., vol. 73, Aug. 1, 1968, p. 4869.
18. FORMISANO, V.: Interplanetary Plasma Electrons—A Preliminary Report of Pioneer 6 Data. J. Geophys. Res., vol. 74, Jan. 1, 1969, p. 355.
19. LAZARUS, A. J.; AND BINSACK, J. H.: Observations of the Interplanetary Plasma (Associated with the July 7, 1966, Flare). Paper presented at 10th COSPAR Meeting, 1967.
20. HOWE, H. C., JR.: Pioneer 6 Plasma Measurements in the Magnetosheath. J. Geophys. Res., vol. 75, May 1, 1970.
21. LAZARUS, A. J.; SISCOE, G. L.; AND NESS, N. F.: Plasma and Magnetic Field Observations during the Magnetosphere Passage of Pioneer 7. J. Geophys. Res., vol. 73, Apr. 1968.
22. WOLFE, J. H.; ET AL.: Initial Observations of the Interplanetary Solar Wind by the Pioneer 6 Quadrispherical Plasma Probe. Am. Geophys. Union, 1966.
23. INTRILIGATOR, D. S.; ET AL.: Preliminary Comparison of Solar Wind Plasma Observations in the Geomagnetic Wake at 1000 and 500 Earth Radii. Plan. and Space Sci., vol. 17, Mar. 1969.
24. SCARF, F. L.; WOLFE, J. H.; AND SILVA, R. W.: A Plasma Instability Associated with Thermal Anisotropies in the Solar Wind. J. Geophys. Res., vol. 72, Feb. 1, 1967.

25. Fan, C. Y.; et al.: Anisotropy and Fluxuations of Solar Proton Fluxes of Energies 0.6–100 MeV Measured on the Pioneer VI Space Probe. J. Geophys. Res., vol. 71, Jul. 1, 1966.
26. Fan, C. Y.; et al.: Protons Associated with Centers of Solar Activity and Their Propagation in Interplanetary Magnetic Field Regions Corotating with the Sun. J. Geophys. Res., vol. 73, Mar. 1, 1968.
27. Fan, C. Y.; et al.: Differential Energy Spectra and Intensity Variation of 1–20 MeV/Nucleon Protons and Helium Nuclei in Interplanetary Space (1964–1966). Can. J. Phys., vol. 46, May 15, 1968, p. S498.
28. Retzler, J.; and Simpson, J. A.: Relativistic Electrons Confined within the Neutral Sheet of the Geomagnetic Tail. J. Geophys. Res., vol. 74, May 1, 1969, p. 2149.
29. McCracken, K. G.; and Ness, N. F.: The Collimation of Cosmic Rays by the Interplanetary Magnetic Field. J. Geophys. Res., vol. 71, Jul. 1, 1966.
30. Rao, U. R.; McCracken, K. G.; and Bartley, W. C.: Cosmic-Ray Propagation Processes, 3. The Diurnal Anisotropy in the Vicinity of 10 MeV/Nucleon. J. Geophys. Res., vol. 72, Sept. 1, 1967, p. 4343.
31. Balasubrahmanyan, V K.; et al: Co-Rotating Modulations of Cosmic Ray Intensity Detected by Spacecraft Separated in Solar Azimuth. NASA TM–X–63654, 1969.
32. McCracken, K. G.; Rao, U. R.; and Ness, N. F.: Interrelationship of Cosmic-Ray Anisotropies and the Interplanetary Magnetic Field. J. Geophys. Res., vol. 73, Jul. 1, 1968.
33. McCracken, K. G.; Rao, U. R.; and Bukata, R. P.: Recurrent Forbush Decreases Associated with M-Region Magnetic Storms. Phys. Rev. Letters, vol. 17, Oct. 24, 1966, p. 928.
34. McCracken, K. G.; Rao, U. R.; and Bukata, R. P.: Cosmic-Ray Propagation Processes, 1. A Study of the Cosmic-Ray Flare Effect. J. Geophys. Res., vol. 72, Sept. 1, 1967.
35. Rao, U. R.; McCracken, K. G.; and Bukata, R. P.: The Acceleration of Energetic Particle Fluxes in Shock Fronts in Interplanetary Space. Can. J. Phys., vol. 46, 1968, p. S844.
36. Bukata, R. P.; McCracken, K. G.; and Rao, U. R.: A Comparison of the Characteristics of Corotating and Flare-Initiated Forbush Decreases. Can. J. Phys., vol. 46, May 15, 1968, p. S994.
37. Bukata, R. P.; et al.: Pioneer VI Observations of Forbush-Type Modulation Phenomena in the Galactic Alpha Particle Flux. Paper presented at 11th Intl. Conf. on Cosmic Rays (Budapest), 1970.
38. Bukata, R. P.; et al.: Neutron Monitor and Pioneer 6 and 7 Studies of the January 28, 1967, Solar Flare Event. Solar Phys., vol. 10, Nov. 1969, p. 198.
39. Reiff, G. A.: Radio Propagation Experiments with Interplanetary Spacecraft. J. Spacecraft and Rockets, vol. 6, May 1969.
40. McCracken, K. G.; et al.: The Decay Phase of Solar Flare Events. Solar Phys., 1971.
41. Lezniak, J. A.; et al.: Observations on the Abundance of Nitrogen in the Primary Cosmic Radiation. Astrophys. and Space Sci., vol. 5, Sept. 1969.
42. Lezniak, J.; and Webber, W. R.: Observations of Fluorine Nuclei in the Primary Cosmic Radiation Made on the Pioneer 8 Spacecraft. Astrophys. J., vol. 156, May 1969, p. L73.
43. Lezniak, J. A.; von Rosenvinge, T. T.; and Webber, W. R.: The Chemical Composition and Energy Spectra of Cosmic Ray Nuclei with $Z = 3$–30. Acta Physica, vol. 29, Supplement 1, 1970.
44. Beedle, R. E.; et al.: Measurements of the Primary Electron Spectrum in the Energy Range 0.2 MeV to 15 GeV. Acta Physica, vol. 29, Supplement 1, 1970.

45. LEZNIAK, J. A.; WEBBER, W. R.; AND ROCKSTROH, J.: A Comparison of Solar Modulation Effects on Protons, Electrons and Helium Nuclei. Paper presented at 11th Intl. Conf. on Cosmic Rays (Budapest), 1970.
46. STANFORD UNIVERSITY AND STANFORD RESEARCH INSTITUTE: The Interplanetary Electron Number Density from Preliminary Analysis of the Pioneer VI Radio Propagation Experiment. J. Geophys. Res., vol. 71, Jul. 1, 1966, p. 3325.
47. KOEHLER, R. L.: Interplanetary Electron Content Measured Between Earth and Pioneer VI and VII Spacecraft Using Radio Propagation Effects. Stanford Electronic Laboratory SU–SEL–67–051, 1967.
48. KOEHLER, R. L.: Radio Propagation Measurements of Pulsed Plasma Streams from the Sun Using Pioneer Spacecraft. J. Geophys. Res., vol. 73, Aug. 1, 1968.
49. LANDT, J. A.; AND CROFT, T. A.: A Plasma Cloud Following a Solar Wind Shock on 7 July 1966 Measured by Radio Propagation to Pioneer 6. Stanford Electronics Laboratory SU-SEL–70–001, 1970.
50. POMALAZA-DIAZ, J. C.: Measurement of the Lunar Ionosphere by Occultation of the Pioneer VII Spacecraft. Stanford Electronics Laboratory SU–SEL–67–095, 1967.
51. CROFT, T. A.: Patterns of Solar Wind Flow Deduced from Interplanetary Density Measurements Taken during 21 Rotations of the Sun in 1968–70. Stanford University SU–SEL–70–063, 1970.
52. LEVY, G. S.; ET AL.: Pioneer 6—Measurement of Transient Faraday Rotation Phenomena Observed during Solar Occultation. Science, vol. 166, Oct 31, 1969.
53. SCHATTEN, K. H.: Evidence for a Coronal Magnetic Bottle at 10 Solar Radii. NASA-TM-X-63811, 1969.
54. SCARF, F. L.; ET AL.: Initial Results of the Pioneer 8 VLF Electric Field Experiment. J. Geophys. Res., vol. 73, Nov. 1, 1968.
55. SCARF, F. L.; FREDRICKS, R. W.; AND KENNEL, C. F.: AC Electric and Magnetic Fields and Collisionless Shock Structures. Particles and Fields in the Magnetosphere. R. M. McCormac, ed., D. Reidel Publishing Co.; Dordrecht, 1970, p. 102.
56. SCARF, F. L.; ET AL.: AC Fields and Wave-Particle Interactions. Particles and Fields in the Magnetosphere. R. M. McCormac, ed., D. Reidel Publishing Co., Dordrecht, 1970, p. 275.
57. SCARF, F. L.; ET AL.: Pioneer 8 Electric Field Measurements in the Distant Geomagnetic Tail. J. Geophys. Res. vol. 75, June 1, 1970, p. 3167.
58. SISCOE, G. L.; ET AL.: VLF Electric Fields in the Interplanetary Medium: Pioneer 8. TRW Systems 10472–6016–RO–00, 1971.
59. SISCOE, G. L.: ET AL.: Evidence for a Geomagnetic Wake at 500 R_E. J. Geophys. Res., vol. 75, Oct. 1, 1970, p. 5319.
60. SCARF, F. L.; GREEN, I. M.; AND CROOK, G. M.: The Pioneer 9 Electric Field Experiment: Part 1, Near Earth Observations. TRW Systems 10472–6019–RO–00, 1970.
61. SCARF, F. L.; AND SISCOE, G. L.: The Pioneer 9 Electric Field Experiment: Part 2, Observations between 0.75 and 1.0 AU. TRW Systems 10472–6120–RO–00, 1970.
62. BERG, O. E.; KRISHNA SWAMY, K. S.; AND SECRETAN, L.: Cosmic Dust Radiants and Velocities from Pioneer 8. Goddard Space Flight Center X–616–69–145 and X–616–69–233, 1969.
63. BERG, O. E.; AND GERLOFF, U.: More Than Two Years of Micrometeorite Data from Two Pioneer Satellites. Paper presented at XIIIth Plenary Meeting of COSPAR (Leningrad), May 1970.
64. ANDERSON, J. D.; AND HILT, D. E.: Improvement of Astronomical Constants and Ephemerides from Pioneer Radio Tracking Data. Am. Astronaut. Society Paper 68–130, 1968.

Bibliography

ANON.: Significant Achievements in Space Science, 1966. NASA SP-155, 1967.
ANON.: Significant Achievements in Space Science, 1968. NASA SP-167, 1968.
BAITY, W. H.; ET AL.: Low Energy Cosmic Ray H^2 and He^3 Nuclei Intensities Measured in 1968. Astrophys. J., 1970.
BARTLEY, W. C.; ET AL.: Anisotropic Cosmic Radiation Fluxes of Solar Origin. J. Geophys. Res., vol. 71, Jul. 1, 1966, p. 3299.
——; ET AL.: Pioneer 6 Measurements of the Degree of Anisotropy of the Galactic Cosmic Radiation. COSPAR paper, 1966.
BERG, O. E.; AND GERLOFF, U.: Orbital Elements of Micrometeorites Derived from Pioneer 8 Measurements. J. Geophys. Res., vol. 75, Dec. 1, 1970, p. 6932.
BINSACK, J. H.; AND LAZARUS, A. J.: Observations of the Interplanetary Plasma Subsequent to the 7 July Proton Flare. Annals of IQSY, vol. 3, 1969, p. 378.
BUKATA, R. P.; ET AL.: Neutron Monitor and Pioneer 6 and 7 Studies of the January 28, 1967, Solar Flare Event. Solar Phip., vol. 10, Nov. 1969, p. 198.
——: Pioneer 6 and 7 Observations of Solar Induced Cosmic Radiation. Paper presented at Am. Geophys. Union Meeting (Washington), 1967.
——; MCCRACKEN, K. G.; AND RAO, U. R.: Co-rotating Modulation Phenomena. Paper presented at 10th Intl. Conf. on Cosmic Radiation (Calgary), 1967.
——; AND PALMEIRA, R. A. R.: The Effect of the Filamentary Interplanetary Magnetic Field Structure on the Solar Flare Event of May 4, 1966. J. Geophys. Res., vol. 72, Nov. 1, 1967, p. 5563.
—— Anisotropic Proton Propagation Observed with Pioneers 6 and 7. Paper presented at Midwest Cosmic Ray Conf., State Univ. of Iowa, 1968.
——: Summary of Low Energy Solar Cosmic Ray Events. Paper presented at the 7th Aerospace Sci. Meeting (New York), 1969.
BURLAGA, L. F.: Hydromagnetic Discontinuities in the Solar Wind. Significant Accomplishments in Science, 1968. NASA SP-195, 1969, pp. 127–130.
——: On the Nature and Origin of Directional Discontinuities. Goddard Space Flight Center X-692-70-462, 1970.
——: Microstructure of the Interplanetary Medium. Goddard Space Flight Center X-692-71-100, 1970.
CANTARANO, S.; AND MARIANI, F.: Magnetic Field Measurements in Interplanetary Space. European Space Res. Org. ESRO-SR-4,
——; NESS, N. F.; AND SCEARCE, C. S.: Preliminary Results from Pioneer VI Magnetic Field Experiment. J. Geophys. Res., vol. 71, Jul. 1, 1966, p. 3305.
CROFT, T. A.; AND HOWARD, H. T.: The Line of Corotating Structures in the Solar Wind Deduced from Electron Content Measured Along the Line of Sight to Four Pioneers. Paper presented at Am. Geophys. Union Meeting (San Francisco), 1969.
——: Corotating Regions in the Solar Wind, Evident in Number Density Measured by Radio Propagation Technique. Radio Sci., vol. 6, Jan. 1971, p. 55.
DHANJU, M. S.; AND SARABHAI, V.: Short Period Fluctuations of Cosmic Ray Intensity at Geomagnetic Equator and Their Solar and Terrestrial Relationships. J. Geophys. Res., vol. 75, Apr. 1, 1970, p. 1795.

DRYER, M.; AND JONES, D. L.: Energy Deposition in the Solar Wind by Flare-Generated Shock Waves. J. Geophys. Res., vol. 73, Aug. 1, 1968, p. 4875.

ESHLEMAN, V. R.; ET AL.: Pioneer VIII Lunar Occultation. URSI Paper (Ottawa), 1967.

———: Bistatic Radar Astronomy. Am. Astronaut. Society Paper 68-182, 1968.

FAIRFIELD, D. H.: Magnetic Field of the Magnetosphere and Tail. The Polar Ionosphere and Magnetospheric Processes. G. Skovli, ed., Gordon and Breach (New York), 1970, pp. 1-23.

FISK, L. A.; AND AXFORD, W. I.: Radial Gradients and Anisotropies of Cosmic Rays. NASA TM-X-63743, 1969.

FORMAN, M. A.: The "Equilibrium" Anisotropy in the Flux of 10 MeV Solar Flare Particles and Their Convection in the Solar Wind. J. Geophys. Res., vol. 75, June 1, 1970, p. 3147.

GERLOFF, U.; AND BERG, O. E.: A Model for Predicting the Results of "In Situ" Meteoroid Experiments, Part 1: Pioneer 8 and 9 Results and Phenomenological Evidences. Paper presented at the XIIIth Plenary Meeting of COSPAR (Leningrad), May 1970.

GOLDSTEIN, R. M.: Pioneer 6: Measurement of Transient Faraday Rotation Phenomena Observed during Solar Occultation. Science, vol. 166, Oct. 31, 1969, p. 596

———: Superior Conjunction of Pioneer 6: Science, vol. 166, Oct. 31, 1969, p. 598.

GREEN, I. M.; AND SCARF, F. L.: Description of Full Data Plots for the Pioneer 8 Electric Field Experiment. TRW Systems 10472-6005-RO-00, 1969.

———: Pioneer 9 Solar Wind Velocity and Electric Field Data, 8 Nov. 1968 to 4 Jan. 1969. World Data Center Report, UAG-9, 1970, p. 40.

GRINGAUZ, K. I.; ET AL.: Comparison of the Simultaneous Measurements of the Concentration and Velocity of Solar-Wind Ions by the Venera-3 and Pioneer 6 Space Vehicles. Kosmicheskie Issledovaniia, vol. 5, May 4, 1967, p. 310. (Also available as NASA CR-84086, 1967.)

HOWARD, H. T.; AND KOEHLER, R. L.: Interplanetary Electron Concentration and Variability Measurements with Pioneer VI and VII. The Zodiacal Light and the Interplanetary Medium. J. L. Weinberg, ed., NASA SP-150, 1967, pp. 361-364.

HUNDHAUSEN, A J.: Interpretation of Positive Ion Measurements Made by the ARC Quadrispherical Electrostatic Analyzer on Pioneer 6. Preprint, no date.

KENNEL, C. F.; AND SCARF, F. L.: Thermal Anisotropies and Electromagnetic Instabilities in the Solar Wind. J. Geophys. Res., vol. 73, Oct. 1, 1968, p. 6145.

KHOKHLOR, M. Z.: Interpretation of the Data on the Angular Distribution of Solar Wind Ions Obtained on Pioneer 6. Geomagnetism and Aeronomy, vol. 7, 1968, p. 874.

LANDT, J. A.; AND ESHLEMAN, V. R.: An Estimate of the Spatial Extent of a Flare-Induced Plasma Cloud Using Columnar Electron Content. Paper presented at Am. Geophys. Union Meeting (San Francisco), 1969.

———; AND CROFT, T. A.: Shape of a Solar-Wind Disturbance on 9 July 1966 Inferred from Radio-Signal Delay to Pioneer 6. J. Geophys. Res., vol. 75, Sept. 1, 1970, p. 4623.

LEVY, G. S.; ET AL.: Pioneer VI Faraday-Rotation Solar Occultation Experiment. Paper presented at 12th COSPAR Meeting, 1969.

LEZNIAK, J. A.: A Measurement of Charged Particle Spectra in Interplanetary Space. Ph.D. Thesis, Univ. of Minnesota, 1969.

———; AND WEBBER, W. R.: Measurements of Gradients and Anisotropies of Cosmic Rays in Interplanetary Space. Paper presented at 11th Intl. Conf. on Cosmic Rays (Budapest), 1970.

———: Solar Modulation of Cosmic Rays, Protons, Helium Nuclei, and Electrons. J. Geophys. Res., vol. 76, Mar. 1, 1971, p. 1605.

LOCKWOOD, J. A.; AND WEBBER, W. R.: Cosmic Ray Intensity Variations on January 26-27, 1968. J. Geophys. Res., vol. 74, Nov. 1, 1969, p. 5599.
——; ET AL.: Comparison of the Rigidity Dependence for Forbush Decreases in 1968 with that for the 11-Year Variation, J. Geophys. Res., 1970.
LOTOVA, N. A.; AND RUKHADZE, A A.: On the Nature of the Inhomogeneous Structure of Interplanetary Plasma, Astronomicheskiy Zhurnal, vol. 45, 1969, p. 343.
MARIANI, F.: Recent Observations of the Geomagnetic Tail. The Earth Explored and Served by Satellites, 1969. In Italian, AIAA abstract A70-13851.
——; AND NESS, N. F.: Reinterpretation of the Pioneer 6 Bow Shock Crossing. Goddard Space Flight Center X-690-71-24, 1971.
MCCRACKEN, K. G.; ET AL.: A Co-Rotating Solar Cosmic Ray Enhancement Observed by Pioneer 8 and Explorer 34 on July 13, 1966. Paper presented at 11th Intl. Conf. on Cosmic Radiation (Budapest), 1970.
MCDONALD, F. B.: IQSY Observations of Low-Energy Galactic and Solar Cosmic Rays. Annals of the IQSY, vol. 4, 1969, p. 187.
MERRITT, R. P.; AND BARON, M. J.: Ionospheric Integrated Electron Content for Use with Pioneer Interplanetary Spacecraft. URSI Paper (Ottawa), 1967.
MIHALOV, J. D.; SONETT, C. P.; AND WOLFE, J. H.: Hugoniot Equations Applied to Earth's Bow Shock Wave. J. Plasma Phys., vol. 3, Sept. 1969, p. 449.
——; ET AL.: Pressure Balance Across the Distant Magnetopause. Cosmic Electrodynamics, vol. 1, Dec. 1970, p. 389.
NESS, N. F.: Solar Wind and the Interplanetary Magnetic Field. Astronaut. Aeron., vol. 5, Oct. 1967, p. 8.
——; AND WILCOX, J. M.: Interplanetary Sector Structure, 1962-1966. Solar Phys., vol. 2, Nov. 1967, p. 351.
——: Direct Measurements of Interplanetary Magnetic Field and Plasma. Annals of the IQSY, vol. 4, 1969, p. 88.
——: The Magnetic Structure of Interplanetary Space. NASA TM-X-63634, .1969.
——; MARIANI; F.; AND BAVASSANO, B.: Interplanetary Magnetic Field Measured by Pioneer 8 during the 25 February 1969 Event. NASA TM-X-63790, 1969.
——; AND SCHATTEN, K. H.: Detection of Interplanetary Magnetic Field Fluctuations Stimulated by the Lunar Wake. J. Geophys. Res., vol. 74, Dec. 1, 1969, p: 6425.
NUNAMAKER, R. R.; HALL, C. F.; AND FROSOLONE, A.: Solar Weather Monitoring—Pioneer Project. Am. Inst. of Astronaut. and Aeron. Paper 68-36, 1968.
PICHLER, H.: Analysis of the Solar Wind Data from Pioneer VI. Archiv fur Meteorologie, Geophysik und Bioklimatologie, Series A, vol. 19, No. 2, 1970, p. 187.
RAO, U. R.; MCCRACKEN, K. G.; AND BUKATA, R. P.: Cosmic-Ray Propagation Processes, 2. The Energetic Storm Particle Event. J. Geophys. Res., vol. 72, Sept. 1, 1967, p. 4325.
——: Pioneer 6 Observations of the Solar Flare Particle Event of 7 July 1966. Annals of the IQSY, vol. 3, 1967, p. 329.
REIFF, G. A.: Summary Results from Interplanetary Spacecraft Radio Propagation Experiments. NASA, no date.
SARI, J. W.: Power Spectral Studies of the Interplanetary Magnetic Field. Acta Physica, vol. 29, supp. 2, 1970, p. 373.
SCARF, F. L.: In-Orbit Interference Problems. JPL Electro-Magnetic Interference Workshop, NASA CR-100697, 1968, pp. 149-162.
——; AND FREDRICKS, R. W.: Ion Cyclotron Whistlers in the Solar Wind. J. Geophys. Res., vol. 73, Mar. 1, 1968, p. 1747.
——: Interplanetary Waves and Their Effects on the Magnetosphere. TRW Systems 10472-6009-RO-00, 1969.
——; ET AL.: Observations of Plasma Waves in Space. Proceedings of the NATO Ad-

vanced Study Institute's Plasma Waves in Space and in the Laboratory, 1969. J. O. Thomas and B. J. Landmark, eds., American Elsevier Publishing Co., 1969, pp. 379–404.

——: Analysis of Pioneer C and D EFD Data. Final Report. TRW Systems 10472–6013–RO–00, 1970.

——: Microstructure of the Solar Wind. TRW Systems 10472–6015–RO–00, 1970.

——; ET AL: Magnetic and Electric Field Changes Across the Shock and in the Magnetosheath. TRW Systems 05402–6011–RO–00, 1969. (Also in: Intercorrelated Satellite Observations Related to Solar Events. V. Manno and D. E. Page, eds., D. Reidel Publishing Co. (Dordrecht), 1970, pp. 181–189.)

SCHATTEN, K. H.: Large-Scale Configuration of the Coronal and Interplanetary Magnetic Field. Univ. of Calif. Rept., 1968.

——; NESS, N. F.; AND WILCOX, J. M.: Influence of a Solar Active Region on the Interplanetary Magnetic Field. Solar Phys., vol. 5, 1968, p. 240.

——: Large-Scale Properties of the Interplanetary Magnetic Field. Goddard Space Flight Center X-692-71-96, 1971.

STANFORD UNIVERSITY AND STANFORD RESEARCH INSTITUTE: Interplanetary and Terrestrial Wake Electron Number Density Measurements with Pioneer 6 and 7. 1966.

STELZRIED, C. T.: A Faraday Rotation Measurement of a 13 cm Signal in the Solar Corona. Ph.D. Thesis, Univ. of Southern California, 1969.

WILCOX, J. M.; AND NESS, N. F.: Quasi-Stationary Corotating Structure in the Interplanetary Medium. J. Geophys. Res., vol. 70, Dec. 1, 1965, p. 5793.

——: The Interplanetary Magnetic Field—Solar Origin and Terrestrial Effects. Space Sci. Rev., vol. 8, Apr. 1968, p. 258.

WOLFE, J. H.; ET AL.: The Compositional, Anisotropic, and Nonradial Flow Characteristics of the Solar Wind. J. Geophys. Res., vol. 71, Jul. 1, 1966, p. 3329.

——; ET AL.: Preliminary Observations of a Geomagnetospheric Wake at 1000 Earth Radii. J. Geophys. Res., vol. 72, Sept. 1967, p. 4577.

——; ET AL.: Preliminary Pioneer 8 Observations of the Magnetospheric Wake at 500 Earth Radii. Trans. of the Am. Geophys. Union, vol. 49, 1968, p. 517 (abstract only).

——; AND MCKIBBIN, D.D.: Pioneer 6 Observations of a Steady-State Magnetosheath. Planetary and Space Sci., vol. 16, 1968, p. 953.

——: Review of Ames Research Center Plasma-Probe Results from Pioneers 6 and 7. Physics of the Magnetosphere. R. L. Carovillano, J. F. McClay, and H. R. Rodoski, eds., D. Reidel Publishing Co. (Dordrecht), 1968, pp. 435–460.

Index

Ames magnetometer, 23, 127
Ames plasma probe, 127
  scientific results, 90, 92–97, 130
Ames Research Center, 1, 2, 6, 7, 9, 13, 55, 133
  (*See also* Ames magnetometer, Ames plasma probe)
California Institute of Technology, 124
Cape Canaveral (Cape Kennedy), 3, 4, 5, 13, 37
Celestial mechanics experiment results, 138, 140–141
Chicago cosmic-ray experiment, 20
  scientific results, 97–102
Communication subsystem performance, 64–65, 66, 69, 70, 71
Convolutional coder, 23
  performance, 74–75
Cosmic dust, 135–138, 139, 140
  (*See also* Goddard cosmic-dust experiment)
Cosmic rays, 97–115
  (*See also* Chicago cosmic-ray experiment, GRCSW cosmic-ray experiment, Minnesota cosmic-ray experiment)
Data-handling subsystem performance, 65, 67–68, 73
Deep Space Network (DSN), 3, 4, 6, 7, 9, 25, 115, 124
  launch configuration, 37
  (*See also* Tracking and data acquisition)
Delta launch vehicle, 3, 4, 11, 17, 25
  performance, 26–36
Department of Defense, 142
DSN. *See* Deep Space Network.
Earth, magnetic field, 78–83, 86–87, 88–91, 92, 95–97, 102, 129–131
Eastern Test Range (ETR), 3, 4, 8–9, 11, 17, 25, 36, 37, 38, 39

Electrical Ground Support Equipment (EGSE), 3, 10
Electric-power subsystem performance, 60, 62, 63, 71–72
Environmental Science Service Administration (ESSA), 142
Goddard cosmic-dust experiment, scientific results, 135–138, 139, 140
Goddard magnetometer, 72, 103, 127
  scientific results, 78–85, 86, 103, 133, 134
Goddard Space Flight Center (GSFC), 2, 8, 9, 127
GRCSW cosmic-ray experiment, 78
  scientific results, 102–110
Ground Operational Equipment (GOE), 4, 6, 7
International Quiet Sun Year (IQSY), 77
Jet Propulsion Laboratory (JPL), 1, 2, 6–7, 38, 39
  (*See also* Deep Space Network, JPL celestial mechanics experiment)
JPL celestial mechanics experiment, scientific results, 138, 140–141
Launch sequence, 27, 30
Launch vehicle. *See* Delta launch vehicle.
Magnetometers. *See* Goddard magnetometer.
Manned Space Flight Network (MSFN), 25, 37, 38
McDonnell-Douglas Astronautics Co., 2, 8
Micrometeoroids. *See* Cosmic dust.
MIT Faraday-cup plasma probe, 72
  scientific results, 85–90, 119
NASA Communications Network (NASCOM), 11
NASA Headquarters, 8, 142

Orientation, Type I maneuver, 25, 41, 44, 45, 46–49
  (*See also* specific spacecraft, flight operations)
  Type II maneuver, 7, 11, 12, 46–49
  (*See also* specific spacecraft, flight operations)
Orientation subsystem performance, 58, 59, 60, 69
  (*See also* Sun-sensor degradation)
Pioneer A. *See* Pioneer 6.
Pioneer B. *See* Pioneer 7.
Pioneer C. *See* Pioneer 8.
Pioneer D. *See* Pioneer 9.
Pioneer E, 12, 22, 26
  flight operations, 43–44
  launch-vehicle performance, 30, 36
  prelaunch narrative, 23
Pioneer 6, 21
  flight operations, 38, 40, 41, 49–51
  gas leak, 58, 59
  launch-vehicle performance, 26–28, 29, 30
  prelaunch narrative, 20
  spacecraft performance, 55–56, 58–68, 70
  trajectory, 32
Pioneer 7, flight operations, 41, 43, 51
  launch-vehicle performance, 28, 30
  prelaunch narrative, 20–21
  spacecraft performance, 56–57, 68–69, 71–73
  trajectory, 33
Pioneer 8, 14
  flight operations, 41–52
  launch-vehicle performance, 28, 30
  prelaunch activities, 18–19
  prelaunch narrative, 21–22
  spacecraft performance, 57, 73–74
  trajectory, 34
Pioneer 9, countdown schedule, 16–17
  detailed task sequence, 13, 15–16
  flight operations, 41, 43, 52–53
  launch-vehicle performance, 30, 36
  prelaunch narrative, 23
  spacecraft performance, 57, 74–75
  trajectory, 35
Plasma probes. *See* Ames plasma probe, MIT Faraday-cup plasma probe.
Radio propagation experiments, scientific results, 124
  (*See also* Stanford radio propagation experiment)
SFOF. *See* Space Flight Operations Facility.
Solar weather monitoring, 141–142
Space Flight Operations Facility (SFOF), 1, 3, 4, 6, 7, 11, 13, 38, 45
Stanford radio propagation experiment, scientific results, 115, 118–124, 125
Structure subsystem performance, 65, 72
Sun, magnetic field, 78–85, 92, 95–97, 100, 103–105
  plasma, 78, 81, 83, 85–90, 92–97, 100, 119–124, 129–135
  solar weather, 141–142
Sun-sensor degradation, 58, 69, 71, 73–74
Test and Training Satellite (TETR), 9, 21, 44
Thermal control subsystem performance, 58–61, 72, 73
Tracking and data acquisition, launch to DSS acquisition, 36–44
  launch trajectories, 32–35
  Pioneer-6 ground track, 29
  (*See also* Deep Space Network)
TRW Systems, 2, 7, 9, 55, 133
TRW Systems electric field detector, 78
  scientific results, 90, 127–135
University of Southern California, 124
U.S. Air Force, 2
  (*See also* Eastern Test Range)

www.ingramcontent.com/pod-product-compliance
Lightning Source LLC
Chambersburg PA
CBHW051807170526
45167CB00005B/1908